Microbiologically Influenced Corrosion

Microbiologically Influenced Corrosion

Brenda J. Little
Naval Research Laboratory
Stennis Space Center, MS

Jason S. Lee
Naval Research Laboratory
Stennis Space Center, MS

WILEY-INTERSCIENCE
A JOHN WILEY & SONS, INC., PUBLICATION

For general information on our other products and services or for technical support, please contact our Customer Care Department within the United States at (800) 762-2974, outside the United States at (317) 572-3993 or fax (317) 572-4002.

Wiley also publishes its books in a variety of electronic formats. Some content that appears in print may not be available in electronic formats. For more information about Wiley products, visit our web site at www.wiley.com.

Wiley Bicentennial Logo: Richard J. Pacifico

Library of Congress Cataloging-in-Publication Data:

Little, Brenda, 1945–
  Microbiologically influenced corrosion/Brenda Little, Jason S. Lee.
    p. cm.
  ISBN 978-0-471-77276-7
1. Microbiologically influenced corrosion. 2. Materials—Microbiology. I. Lee, Jason S. II. Title.
TA418.74.L547 2007
620.1'1223—dc22

                                                                      2006027643

10 9 8 7 6 5 4 3 2 1

Contents

9. Microbiologically Influenced Corrosion of Nonmetallics   217

10 Strategies to Prevent or Mitigate Microbiologically Influenced Corrosion   237

Preface

It is extremely difficult to determine the economic costs associated with microbiologically influenced corrosion (MIC), that is, corrosion resulting from the presence and activities of microorganisms. However, MIC has been documented for metals exposed to seawater, freshwater, distilled and demineralized water, process chemicals, foodstuffs, soil, oil, gasoline and aircraft fuels, human plasma, and sewage. Microbiologically influenced corrosion has been documented in chemical, food, and pulp and paper processing; conventional and nuclear power generation; exploration, production, transportation, storage, and use of hydrocarbon fuels; and marine industries and fire protection systems. Microbiologically influenced corrosion is also recognized as a potential problem in long-term nuclear waste storage. In addition, the term MIC has been expanded to include biodegradation of metal matrix and polymeric composites, polymers and paints, ceramics, glass, sandstone, heritage stone works, sealants, caulks, concrete, and other nonmetallics.

Microbiologically influenced corrosion does not produce a unique type of corrosion. Most MIC is localized corrosion and can take the form of pitting, crevice corrosion, underdeposit corrosion, and de-alloying, in addition to enhanced galvanic and erosion corrosion. There are numerous mechanisms and causative organisms for MIC that can vary among metals and alloys and operating conditions for the same materials. Within the past decade, there have been several important discoveries such as the following:

(1) Microbiologically influenced corrosion occurs in environments where no corrosion is predicted.

(2) Corrosion rates resulting from MIC can be extraordinarily fast.

(3) Liquid culture techniques do not provide an accurate assessment of the numbers and types of microorganisms from the natural environment.

(4) The chemical composition of the electrolyte influences not only the numbers and types of microorganisms, but also their impact on corrosion.

(5) Mitigation and control strategies have shifted from the use of biocides to manipulation of the environment, for example, removal of sulfate and addition of nitrate to control specific microbial populations.

The study of MIC has matured into an interdisciplinary science, including electrochemical, metallurgical, surface analytical, microbiological, biotechnological, and biophysical analyses. As a result, fundamental information related to MIC has been published in journals and symposia proceedings for several disciplines. The goal of this text has been to collect material from these diverse sources and present it in a comprehensive, coherent format with references for additional reading. The chapters are arranged so that biofilm formation is introduced as a general topic. The chapter following on causative organisms is more specific, relating bacteria and fungi to corrosion mechanisms for groups of metals. A chapter on electrochemical techniques provides an overview of methods for detection of MIC and elucidation of mechanisms. Chapters on diagnosing and monitoring do not outline cookbook approaches, but are intended to provide the reader with a discussion of possibilities and limitations. Individual chapters are devoted to the impact of alloying elements, including antimicrobial metals, and design features on MIC. Case histories are presented for both general environments, that is, subterranean, atmospheric, and marine exposures, in addition to specific environments and industries. A chapter has been included on MIC of nonmetallics. The concluding chapter is a discussion of strategies for control or prevention of MIC. Control and mitigation of MIC have not been reduced to standard practice. Proposed mitigation strategies include engineering, chemical, and biological approaches.

This volume is intended for the reader who has a basic understanding of corrosion processes, including nomenclature and vocabulary, and who is interested in the impact of microorganisms on those processes. It is expected that this text will be useful to those requiring an introduction to the subject of MIC and also to those expecting a comprehensive treatment of the subject.

BRENDA J. LITTLE
JASON S. LEE

Chapter 1

Biofilm Formation

INTRODUCTION

The term microbiologically influenced corrosion (MIC) is used to designate corrosion due to the presence and activities of microorganisms, that is, those organisms that cannot be seen individually with the unaided human eye, including microalgae, bacteria, and fungi. Microorganisms can accelerate rates of partial reactions in corrosion processes or shift the mechanism for corrosion. Microorganisms do not produce unique types of corrosion; instead, they produce localized attack including pitting, dealloying, enhanced erosion corrosion, enhanced galvanic corrosion, stress corrosion cracking, and hydrogen embrittlement. Microbiologically influenced corrosion has been reported for all engineering metals and alloys with the exception of predominantly titanium and high chromium–nickel alloys, and has been documented for metals and nonmetals exposed to seawater, freshwater, distilled/demineralized water, crude and distillate hydrocarbon fuels, process chemicals, foodstuffs, soils, human plasma, saliva, and sewage. It occurs in environments where corrosion would not be predicted (e.g., low chloride waters) and the rates can be exceptionally high. According to a recent survey, damage due to corrosion in the United States is estimated at \$276 billion. Similar surveys in the United Kingdom, Japan, Australia, and Germany estimate the cost of corrosion to be 1 to 5 percent of the gross national product (www.corrosion-doctors.org). Microbiologically influenced corrosion is reported to account for 50 percent of the total cost of corrosion (Fleming, 1996). The industries most affected by MIC are power generation; oil production, transportation, and storage; and water distribution.

BIOLOGICALLY ACTIVE ENVIRONMENTS

Microorganisms require water, nutrients, and electron acceptors. Liquid water is needed for all forms of life and the availability of water influences the distribution and growth of microorganisms. Water availability can be expressed as equilibrium

Microbiologically Influenced Corrosion By Brenda J. Little and Jason S. Lee
Published 2007 by John Wiley & Sons, Inc.

relative humidity or water activity (a_w) with values ranging from 0 to 1. Microbial growth has been documented over a range of a_w from 0.60 to 0.999, though none can grow at $a_w = 1$ (pure water) because there are no nutrients available to the organism. A representation of the relative requirements of the major elements required for typical microorganisms composition is as follows: $C_{169}(H_{280}O_{80})N_{30}P_2S$. Waters with suitable forms of carbon, nitrogen, phosphorus, and sulfur support microbial growth. Microorganisms can use a variety of electron acceptors for respiration, including oxygen, sulfate, nitrate, nitrite, carbon dioxide, Fe^{3+}, Mn^{4+}, and Cr^{6+}. The significance of electron acceptors will be discussed later in this chapter.

Microorganisms include bacteria, fungi, and microalgae. Algae are unicellular photosynthetic organisms found in a wide range of environments—from freshwater to concentrated brines (pH from 5.5 to 9.0) and temperatures from below 0 to 40 °C. In the presence of light, algae produce oxygen (photosynthesis). In the absence of light, algae consume oxygen (respiration) and reverse the process. Diatoms are microalgae that have silicon-containing frustules and are often the most conspicuous constituents within the biofilm (Figure 1-1a, b). Some diatoms can grow nonphotosynthetically. Many algae excrete organic acids and are primary producers of nutrients that are necessary to support other fouling species. Fungi are nonphotosynthetic, having a vegetative structure known as a mycelium that is the outgrowth of a single reproductive cell or spore (Figure 1-2a, b). Neither spores nor mycelia are capable of movement. Fungi often reach macroscopic dimensions due to mycelia growth. Fungi assimilate organic material and produce organic acids including oxalic, lactic, acetic, and citric. Yeasts are fungi that multiply by forming buds instead of mycelia. Fungi are the most desiccant-resistant microorganisms and can remain active down to $a_w = 0.60$, whereas few bacteria remain active at a_w values below 0.9.

Bacteria have received the most attention for their influence on corrosion. Bacteria can be subdivided into groups depending on shape (Figure 1-3a–c), requirements for oxygen, source of energy, and type of environment in which they survive.

FIGURE 1-1a   Pennate diatoms embedded in a biofilm. (Image by Richard Ray, Naval Research Laboratory, Stennis Space Center, MS.)

FIGURE 1-1*b* Centric diatom *Coscinodiscus* sp. embedded in a biofilm. (Image by Richard Ray, Naval Research Laboratory, Stennis Space Center, MS.)

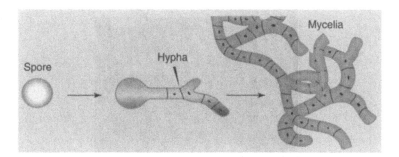

FIGURE 1-2*a* Schematic of fungal spore development into mycelia.

They may occur individually, but tend to form colonies, reproducing by binary fission or cell division. Bacteria range in size from 0.2 μm wide to 1 to 10 μm long. Some filaments may be several hundred millimeters long. Dwarf cells can form in oligotrophic (i.e., nutrient-deprived) waters. Bacteria can be grouped according to their requirements for oxygen and sources of energy. Obligate aerobes require oxygen for survival and growth. Microaerophilic bacteria require low oxygen concentration and facultative anaerobic bacteria can grow under aerobic or anaerobic conditions. Obligate anaerobic microorganisms cannot tolerate oxygen for growth and survival. Obligate anaerobic bacteria are, however, routinely isolated from oxygenated environments associated with particles, crevices, and, most importantly, other bacteria that effectively remove oxygen from the immediate vicinity of the anaerobe. In aerobic respiration, energy is derived when electrons are transferred to oxygen, the terminal electron acceptor. In anaerobic respiration, a variety of organic

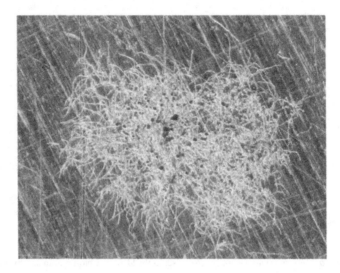

FIGURE 1-2*b* Fungal spores and mycelium on a metal surface. (Image by Richard Ray, Naval Research Laboratory, Stennis Space Center, MS.)

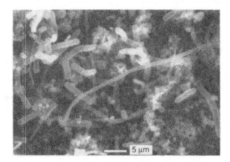

FIGURE 1-3*a* Filamentous and rod-shaped bacteria. (Image by Richard Ray, Naval Research Laboratory, Stennis Space Center, MS.)

FIGURE 1-3*b* Spherical and rod-shaped bacteria. (Image by Richard Ray, Naval Research Laboratory, Stennis Space Center, MS.)

FIGURE 1-3c Spirillum-shaped bacteria. (Image by Richard Ray, Naval Research Laboratory, Stennis Space Center, MS.)

TABLE 1-1 Types of Respiration and Examples of Electron Acceptors

Electron acceptor	Product(s)	Organisms
Aerobic respiration		
O_2	H_2O	Strictly aerobic or facultative anaerobic organisms
Anaerobic respiration		
NO_3^-	NO_2^-, N_2O, N_2	Denitrifying bacteria
S^{2-}	SO_4^{2-}	Obligate anaerobe bacteria, sulfate-reducing bacteria
S	S^{2-}	Facultative and obligate anaerobic bacteria
CO_2	Acetate	Acetogenic bacteria
	Methane	Methanogenic bacteria
Fe^{3+}, Mn^{4+}, Cr^{6+}	Fe^{2+}, Mn^{2+}, Cr^{3+}	Metal-reducing bacteria

and inorganic compounds may be used as the terminal electron acceptor (Table 1-1). Bacteria are grouped based on the terminal electron acceptor in anaerobic respiration, for example, sulfate- and metal-reducing bacteria.

Bacteria can also be grouped according to their nutritional requirements. Heterotrophic bacteria derive energy from a wide range of organic molecules. As a group, heterotrophic bacteria can assimilate almost any available carbon molecule, from simple alcohols and sugars to complex polymers. Many heterotrophic microorganisms can grow on trace nutrients in natural waters or distilled water. Bacteria can adapt to a variety of nutrient sources. For example, *Pseudomonas fluorescens* can use over 100 different compounds, including sugars, lipids, alcohols, phenols, and organic acids, as sole sources of carbon and energy. Diversity among organic substrates is peripheral to the issue of energy-generating and energy-conserving processes. Autotrophic bacteria oxidize inorganic compounds, elements, or ions (e.g., NH_3, NO_2^-, CH_4, H_2, SO_4^{2-}, Fe^{2+}, Mn^{2+}) as sources of energy. When both autotrophic and heterotrophic mechanisms operate simultaneously, the metabolism is mixotrophy. Phototrophic bacteria can use light as a source of energy.

The temperature range in which living organisms can grow is that in which liquid water can exist, approximately 0 to 100 °C. Microbial life is possible over a range of 10 pH units or more. Many microorganisms can withstand a 100-fold or greater variations in pressure. Pressure in the depths of the sea is only mildly inhibitory to the growth of many microorganisms. Heavy-metal concentrations as low as 10^{-8} M can inhibit the growth of some microorganisms, while others may be resistant to concentrations of 1,000,000-fold or greater. Microbial species show 1000-fold differences in susceptibility to ultraviolet, beta, and gamma irradiation. The following general statements about microorganisms are taken from Pope (1986):

- Individual microorganisms are small [from less than two-tenths to several hundred micrometers (μm) in length by up to 2 or 3 μm in width], a quality that allows them to penetrate crevices, and some other small spaces easily. Bacterial and fungal colonies can grow to macroscopic proportions.

- Bacteria may be motile and capable of migrating to more favorable conditions or away from less favorable conditions; for example, toward nutrients or away from toxic materials.

- Bacteria have specific receptors for certain chemicals, which allow them to seek out higher concentrations of nutrients.

- Bacteria and fungi can reproduce very quickly (generation times of 18 min have been reported).

- Individual cells can be widely and quickly dispersed by wind, water, animals, or aircraft.

- Microorganisms are resistant to many chemicals (antibiotics, disinfectants, etc.) by virtue of their ability to degrade them or by being impermeable to them [due to extracellular polymeric substances (discussed later), their cell walls, or their cell membrane characteristics]. Resistance may be acquired through mutation or acquisition of a plasmid by naturally occurring genetic exchange between cells.

- Microorganisms have developed several strategies for survival in natural environments: (1) spore formation, (2) biofilm formation, (3) dwarf cells, and (4) a viable, but nonculturable state. Many bacteria and fungi produce spores that are very resistant to temperature (some even resist boiling for over one hour), acids, alcohols, disinfectants, drying, freezing, and many other adverse conditions. Spores may remain viable for hundreds of years and germinate on finding favorable conditions. In the natural environment, there is a difference between survival and growth. Microorganisms can withstand long periods of starvation and desiccation. If conditions are alternately wet and dry, microbes may survive dry periods and will grow during wet periods.

BIOFILM FORMATION

It is convenient and informative to discuss the characteristics of individual groups of microorganisms; however, in natural environments microorganisms form synergistic communities that conduct combined processes which individual species cannot. The term biofilm embraces an enormous range of microbial associations generally found at phase boundaries (Wimpenny, 1996). In aquatic environments, microbial cells attach to solids, including metals. Immobilized cells grow, reproduce, and produce extracellular polymers forming a biofilm. Biofilm accumulation is the net result of attachment, growth, and detachment (Figure 1-4).

Biofilm formation consists of a sequence of steps and begins with adsorption of macromolecules (proteins, polysaccharides, and humic acids) and smaller molecules (fatty acids and lipids) at surfaces. Adsorbed molecules form conditioning films that alter physiochemical characteristics of the interface, including surface hydrophobicity and electrical charge. The amount of adsorbed organic material is a function of ionic strength and can be enhanced on metal surfaces by polarization.

Because of the complexity of microbial binding to surfaces, the terms attachment and detachment are frequently used without referring to specific physical processes. Attachment is due to microbial transport and subsequent binding to surfaces. The extent of bacterial adhesion and the adhesion pattern depend on bacterial characteristics, including cell-surface hydrophobicity and charge, cell size, presence of flagella and pili, and properties of the substratum such as chemical composition, surface roughness, crevices, inclusions, and coverage by oxide films or organic coatings, the composition and strength of the aqueous medium, and the hydraulic flow regime.

During initial stages of biofilm formation, the major factor controlling the rate of colonization is hydrodynamics. Microbial colonization begins with transport of

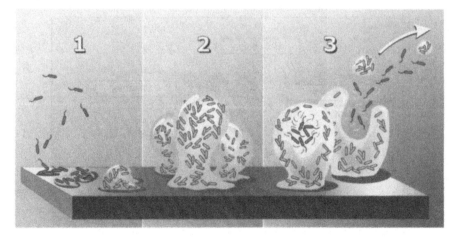

FIGURE 1-4   The biofilm life cycle in three steps: (1) attachment, (2) growth of colonies, and (3) detachment in clumps or "seeding dispersal." (Stoodley and Dirckx, 2003a. Reprinted with permission from the Center for Biofilm Engineering at Montana State University, Bozeman.)

microorganisms to the interface mediated by at least three mechanisms: (1) diffusive transport due to Brownian motion, (2) convective transport due to the liquid flow, and (3) active movement of motile bacteria near the interface. The influence of convection transport exceeds the other two by several orders of magnitude. Once the microbial cell is in contact with a surface, it may or may not adhere. The ratio of cell numbers adhering to a surface to the cell numbers transported to this surface depends on surface properties, physiological state of organisms, and hydrodynamics near the surface. Biofilms grown at high shear stress develop elongated microcolonies (Figure 1-5) (Lewandowski and Stoodley, 1995).

Dense biofilms form as a result of high shear stress or starvation (Beyenal and Lewandowski, 2002). van Loosdrecht et al. (1995) discussed the effects of substrate loading, shear stress, and growth rate on biofilm structure. Immediately after attachment, microorganisms initiate production of slimy adhesive substances, termed extracellular polymeric substances (EPS), which assist in the formation of microcolonies and microbial films. Extracellular polymeric substances bridge microbial cells with the substratum and permit negatively charged bacteria to adhere to both negatively and positively charged surfaces. They may also control interfacial chemistry at the substratum–biofilm interface. Azeredo and Oliveira (2000a) examined the role of exopolymers in biofilm formation and composition. They reported that exopolymers are essential for cell-to-cell adhesion and for biofilm formation. Exopolymers are also responsible for cohesive forces within biofilms and biofilm stability.

Biofilm accumulation at surfaces is an autocatalytic process. Initial colonization increases surface irregularity and promotes further biofilm formation. Increased surface irregularity due to biofilm formation can influence particle transport and attachment rate by (1) increasing convective mass transport near the surface, (2) providing

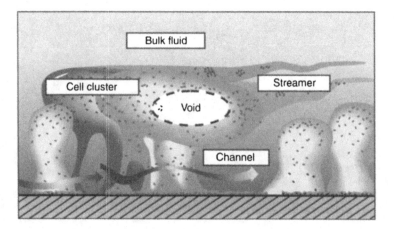

FIGURE 1-5 Conceptual illustration of the heterogeneity of biofilm structure, with labeled bacterial clusters, streamers, and water channels. (Lewandowski and Dirckx, 1996. Reprinted with permission from the Center for Biofilm Engineering at Montana State University, Bozeman.)

shelter from shear forces, and (3) increasing surface area for attachment. Growth is due to microbial replication and growth rate is traditionally described by Monod kinetics:

$$\mu = \frac{\mu_{max} S}{K_s + S}$$

where μ_{max} is the maximum specific growth rate (t^{-1}), K_s the half-saturation coefficient (mol L^{-3}), and S the substrate concentration (mol L^{-3}). Each species in the biofilm has its own optimum growth parameters.

Hydrodynamic shear stress, related to flow, influences transport, transfer, and reaction rates within the biofilm, as well as detachment. Detachment includes two processes: erosion and sloughing. Sloughing is the process in which large pieces of biofilm are rapidly removed, frequently exposing the surface. The reasons for biofilm sloughing are not well understood. Biofilm erosion is defined as continuous removal of single cells or small groups of cells from the biofilm surface and is related to shear stress at the biofilm–fluid interface. Frequent detachment is identified with erosion, especially in conduits. An increase in shear stress increases erosion rate and decreases biofilm accumulation rate. Empirical observations indicate that the erosion rate is related to biofilm thickness and density.

Influence of Conditioning Films

Many investigators have demonstrated that materials with diverse surface properties (e.g., wettability, surface tension, and surface charge) are rapidly conditioned by absorbing organics when exposed to natural waters. The impact of conditioning films on subsequent microbial attachment and growth has been the subject of extensive research and controversy. Poleunis et al. (2002) used surface analytical techniques to monitor the chemical composition and growth kinetics of the adsorbed layer on 316L stainless steel (UNS S31603) immediately after immersion in natural seawater. They reported successive adsorption of two types of compounds before any bacterial attachment—nitrogen-containing species (conjectured to be proteins) followed by carbohydrates. They monitored an increase in adsorbed material for 24 h. Even after a 24-h immersion, there was no continuous conditioning film on the surface. The authors indicated that even in the presence of adsorbed potential nutrients, the substratum influences are more important to bacterial adhesion than the conditioning film. Bradshaw et al. (1997) evaluated the influence of conditioning films on biofilm development using bacteria isolated from the oral cavity and concluded that conditioning films have a role in the degree and pattern of oral biofilm development. However, Ostuni et al. (2001) demonstrated that there is little or no correlation between adsorption of protein on surfaces and adhesion of bacteria. Busscher et al. (1997) concluded that a 1.5-h adsorbed salivary conditioning film appears to slow deposition of yeasts and some bacteria on silicone rubber. They reported that adhesion to silicone rubber is weaker with a salivary conditioning film compared with the same surfaces without the conditioning film, that is, attached cells are easier to remove from conditioned surfaces.

Influence of the Substratum

The surface to which microorganisms attach, the substratum, plays a major role in biofilm processes during the early stages of biofilm accumulation and may influence the rate of cell accumulation and cell distribution. There is a vast literature investigating the influence of surface composition roughness, wettability, and polarization on the attachment of bacteria. It has been demonstrated that the composition of a metal substratum influences the formation rate and cell distribution of microfouling films in seawater during the first hours of exposure. Gerchakov et al. (1977) demonstrated that initial bacterial attachment is more rapid on glass and 304 stainless-steel (UNS S30400) surfaces compared to 60/40 copper-zinc brass (UNS C28000) and 90/10 copper–nickel (UNS C70600) surfaces (Figure 1-6). Hydrated oxide and hydroxide passivating films on

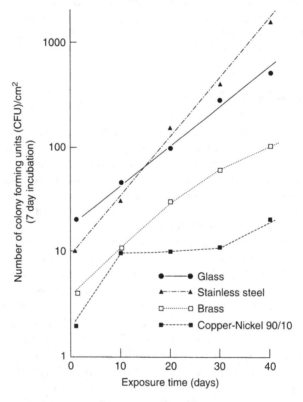

FIGURE 1-6 Numbers of marine heterotrophic bacteria cultured from various substrates in relation to exposure time on surfaces including glass, 304 stainless steel (UNS S30400), 60/40 copper-zinc brass (UNS C28000), and 90/10 copper-nickel (UNS C70600). (Gerchakov et al., 1977. Reprinted with permission of the author.)

metal surfaces provide bacteria with sites for firm attachment (Kennedy et al., 1976). Similarly, spalling or sloughing of corrosion products forces the detachment of the biofilm associated with corrosion products (Characklis et al., 1983).

Conflicting reports have been published regarding the influence of surface topography on microbial attachment. The term roughness is defined as the pattern or texture of surface irregularities that are inherent in the production process, excluding waviness and errors of formation. In some cases, higher surface roughness increased the extent of bacterial accumulation and adhesion took place at surface irregularities, whereas other authors found reduced adhesion to rougher surfaces. Nickels et al. (1981) demonstrated that the microbiota colonizing silica grains of the same size and water pore space but different microtopography show differences in biomass and community structure after 8 weeks of exposure to seawater. Absence of surface cracks and crevices resulted in a marked diminution of total biomass. Medilanski et al. (2002) used four bacterial species comprising three phyla with a variety of physiochemical characteristics to evaluate the influence of surface topography on colonization of UNS S30400. Five types of surface finishes corresponding to roughness values (R_a) between 0.03 and 0.89 μm were produced. Adhesion of all four bacteria was minimal at $R_a = 0.16$ μm, whereas smoother and rougher surfaces gave rise to more adhesion (Figure 1-7).

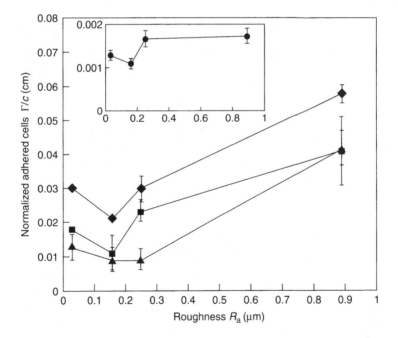

FIGURE 1-7 Normalized levels of adhesion of *Pseudomonas putida* mt2 (■), *Pseudomonas aeruginosa* PAO1 (◆), *Rhodococcus* sp. C125 (▲), and *Desulfovibrio desulfuricans* (●) on UNS S30400 stainless-steel surfaces of different roughness. (Reprinted from Medilanski et al., 2002, with permission from Taylor and Francis Ltd. http://www.tandf.uk/journals.)

The R_a = 0.16 µm surface exhibited parallel scratches 0.7 µm in width, in which a high proportion of bacteria was aligned. This particular surface roughness corresponded to the approximate width of the cells, but was smaller than their length. Bacteria fit these scratches in longitudinal orientation only. The generally higher adhesion to the roughest surfaces may be due to the increased surface area. Reduced adhesion was attributed to unfavorable interactions between the surface and bacteria oriented other than parallel to the scratches. Interaction energy calculations and considerations of microgeometry confirmed this mechanism. Rougher surfaces exhibiting wider scratches allowed a higher fraction of bacteria to adhere in other orientations, whereas the orientation of cells adhered to smoother surfaces was completely random. Flint (1997) suggested that opposing observations regarding the influence of surface roughness on bacteria adhesion are probably related to the degree of surface roughness, the bacterial species tested, the physiochemical parameters of the surface, the bulk fluid phase under study, and the adhesion method used to detect bacteria. Korber et al. (1997) suggested that increased surface area provided at the microorganism–material interface might facilitate more film attachment by providing contact points. Little et al. (1988) demonstrated that porous welds provide increased sites for colonization compared to smooth pipe surfaces. Sreekumari et al. (2001) evaluated bacterial attachment to 304L stainless-steel (UNS S30403) weldments and the significance of substratum microstructure. They found that welded metal samples show more attachment while base metals show the least. The area of attachment was inversely proportional to the average grain size. Bacterial colonization started on grain boundaries. The weld area had more grains and more grain boundaries (Figure 1-8). Furthermore, the authors established a direct relationship between increased attachment and the onset of MIC. There is some evidence that nanometer roughness enhanced the adhesion of the conditioning layer to the substratum (Gold, 1999).

Verran and Boyd (2001) suggested that the surface roughness is also a factor in cell retention on surfaces. If the surface irregularities are much larger than the microorganisms, passive retention is minimal. Surface features on a nanometer scale impact cleanability of surfaces and surface defects on the microbiological scale may confer protection from shear forces in the surrounding environment (Figure 1-9).

Wiencek and Fletcher (1997) used self-assembled monolayers with a range of wettabilities to demonstrate that the greatest number of cells attached to hydrophobic surfaces. Armon et al. (2001) evaluated the impact of polarization on the adsorption of *Flavobacterium breve* (Figure 1-10) and *P. fluorescens* (Figure 1-11) to platinum, titanium (ASTM grade 2, UNS R50400), S31603 stainless-steel, copper (UNS C15000), aluminum alloy (UNS A95052), and carbon steel (UNS G10200). Maximal adsorption occurred in the potential range of -0.5 to 0.5 V (standard calomel electrode, SCE) for all metals. A shift of applied potential toward either the positive or negative direction caused a gradual decrease in bacterial adsorption.

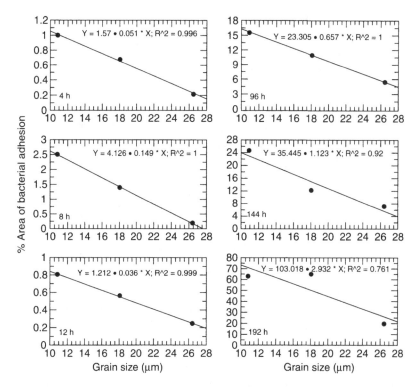

FIGURE 1-8 Results of regression analysis between the average grain size and the percentage areal cover of *Pseudomonas* sp. (Reprinted from Sreekumari et al., 2001, with permission from Taylor and Francis Ltd. http://www.tandf.uk/journals.)

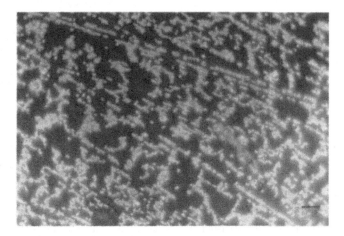

FIGURE 1-9 *Pseudomonas aeruginosa* (stained with acridine orange) retained in surface defects on stainless steel, illustrating the role of surface defects (scratches) in cell retention. Bar = 10 μm. (Adapted from Verran and Boyd, 2001, with permission from Taylor and Francis Ltd. http://www.tandf.uk/journals.)

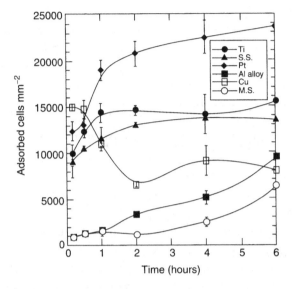

FIGURE 1-10   The effect of applied potential on the adsorption of *P. fluorescens* P17 on six different metal surfaces: platinum (Pt); UNS R50400 (ASTM grade 2) titanium (Ti); UNS 31603 stainless steel
(SS); UNS C15000 copper (Cu); UNS A95052 aluminum alloy (Al); UNS G10200 carbon steel (MS).
(Reprinted from Armon et al., 2001, with permission from Taylor and Francis Ltd.
http://www.tandf.uk/journals.)

Influence of the Electrolyte

Electrolyte concentration, pH, and inorganic ions influence settlement. Fletcher
(1988) found that an increase in the concentration of several cations in the electrolyte
(sodium, calcium, lanthanum, and ferric iron) affects the attachment of *P. fluorescens*
to glass surfaces, presumably by reducing the repulsive forces between the negatively charged bacterial cells and the glass surfaces. Energy derived from organic
carbon drives heterotrophic microbial growth within biofilms. Starvation decreases
adhesion of some species and does not affect others (Wiencek and Fletcher, 1997).
Generally, increasing the total organic carbon (TOC) increases the substrate or carbon source available to the biofilm. Cowan et al. (1991) evaluated the influence of
nutrient concentration on the colonization of glass substrata. They concluded that the
ability of bacteria to colonize surfaces is to a large extent related to their ability to
colonize the liquid phase and deposition of bacteria onto surfaces is positively
correlated with the density of suspended cells. At high organic loadings, the
substrate flux in the biofilm will reach a constant value as a result of one of the following events: (1) the growth rate of the microbial population in the biofilm reaches
a maximum, (2) the thickness of the biofilm exceeds the penetration depth of the
substrate into the biofilm, or (3) the electron acceptor or another nutrient becomes
nutrient-limiting. Others have suggested that attachment to surfaces is a survival

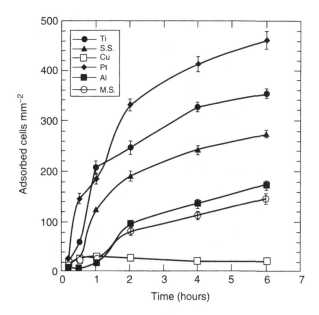

FIGURE 1-11   The effect of applied potential on the adsorption of *F. breve* on six different metal surfaces: platinum (Pt); UNS 31603 stainless steel (SS); UNS R50400 (ASTM grade 2) titanium (Ti); UNS C15000 copper (Cu); UNS G10200 carbon steel (MS); UNS A95052 aluminum alloy (Al). (Reprinted from Armon et al., 2001, with permission from Taylor and Francis Ltd. http://www.tandf.uk/ journals.)

strategy and in oligotrophic waters cells attach to surface-associated organic matter. Organic carbon is present in all natural and processed waters, but its concentration can vary widely. For example, in distilled water TOC is 1 to 2 $g\,m^{-3}$; 1 $g\,m^{-3}$ in the seawater off the coast of Hawaii, 10 $g\,m^{-3}$ in the Gulf of Mexico coastal water, 150 $g\,m^{-3}$ in water produced in oil fields, and 200 $g\,m^{-3}$ in untreated sewage. Biofilm formation has been documented over the widest possible range of substrate concentrations (Table 1-2) (Wimpenny, 1996). Biofilms form in highly oligotrophic waters including ultrapure water. At the other end of the spectrum, biofilms form on surfaces exposed to extremely high nutrient concentrations. Direct evidence indicates that at low nutrients, biofilms form as separate stacks or groups of cells around and through which water can move, and, in high nutrients, biofilms can appear to be dense and almost confluent. Azeredo and Oliveira (2000b) demonstrated that cells entrapped within thin biofilms are metabolically active within a relatively homogeneous matrix. Cells in the inner layer where nutrients are not limited produce a more heterogeneous, denser biofilm structure. In contrast, the inner layer of the thicker biofilm is metabolically inactive. Nutrient limitation in the deeper layer of thick biofilms is responsible for cell lysis and production of proteolytic enzymes.

Carbon is not always the growth-limiting nutrient for microorganisms. Phosphorus and nitrogen may be limiting in some aquatic systems. An electrolyte with a carbon–nitrogen ratio greater than 7:10 is considered nitrogen-limited for

TABLE 1-2   A Brief Survey of Biofilm Types in Relation to Nutrient Availability

Nutrient level	System
Extremely low	High purity water: granular activated carbon, reverse osmosis membrane, ion-exchange resin, degasifier unit, water storage tanks, microporous membrane filters.
	Water distribution and storage: water distribution pipes, domestic copper piping, storage tanks, oligotrophic stream, and river and lake epilithon
Low	Eutrophic water bodies
	Swimming-pool filters
	Domestic drains
	Car wash bottles
	Plant surfaces
	Phylloplane communities
Medium	Effluent treatement: trickling filter, anaerobic digester granules, rotating disc aerators, fluidized-bed reactors
	Production: membrane bioreactor, vinegar production
	Plant surfaces: rhizosphere
High	Food associated: food products (meat, etc.), food processing surfaces
	Animal surfaces: Oral surfaces—cheek, tongue, palate, epithelium, tooth surfaces
	Epithelia—gut, rumen, vagina, etc.
	Infection—lung, heart valves, etc.
	Prosthesis and catheters: pacemaker, metal plates, joints, heart, heart valve stents, indwelling catheters, etc.
	Contact lenses

Source: Reprinted from Wimpenny, 1996, with permission from Taylor and Francis Ltd. http://www.tandf.uk/journals.

microbial growth. Cells growing in such a medium tend to reproduce slowly and produce copious amounts of EPS (Wilkinson, 1958). Biofilm thickness has been shown to increase with increasing carbon–nitrogen ratios (Bott and Gunatillaka, 1983). McEldowney and Fletcher (1988) demonstrated that the carbon source can influence the adhesive qualities of cells.

Bulk water temperature influences the rate of most chemical and biochemical reaction processes as well as transport processes within the biofilm. Biofilm formation is generally considered to be more of a problem in the summer months because higher temperatures increase the rate of biological processes.

SUMMARY

Biofilms form on all engineering materials exposed in biologically active environments. They form compliant surfaces that actively interact with the hydrodynamic boundary layer (Figure 1-12), and can form in extreme environments such as

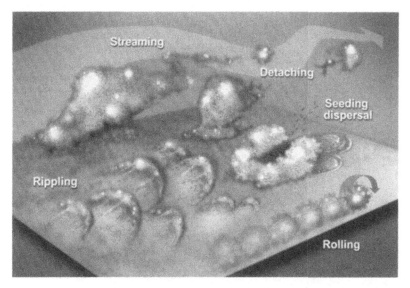

FIGURE 1-12   Biofilm bacteria can move in numerous ways: collectively by rippling or rolling across a surface, or by detaching in clumps or streaming, and individually, through a "swarming and seeding" dispersal. (Stoodley and Dirckx, 2003b. Reprinted with permission from the Center for Biofilm Engineering at Montana State University, Bozeman.)

ultrapure waters (McFeters et al., 1993; Kulakov et al., 2002) or highly radioactive conditions. Lewandowski (1998) hypothesized that biofilm development optimizes survival of the biofilm constituents and maximizes transport of nutrients into the biofilms. Biofilms also provide protective environments for bacteria and in most cases allow different types of bacteria to flourish within different strata of the biofilm (Figure 1-13) (Harrison et al., 2005). For example, obligate anaerobic bacteria are routinely isolated from oxygenated environments in association with other bacteria that effectively remove oxygen from the immediate vicinity of the anaerobe.

Bacteria within the biofilm act symbiotically to produce conditions more favorable for the growth of each species. Bacteria near the fluid phase are provided with complex nutrients and oxygen. These bacteria use oxygen, break down carbon sources, and produce simple polymers and fatty acids. Bacteria within the biofilm, removed from the bulk phase, use waste products generated by other bacteria as nutrients that are metabolized to fatty acids, carbon dioxide, and hydrogen. Cole (1982) reviewed the interaction between bacteria and algae in aquatic systems. Haack and McFeters (1982) demonstrated a flux of dissolved algal products into heterotrophic bacteria. The cycling of carbon and oxygen in algal bacterial aggregates and biofilms has also been observed in the marine environment (Azam and Ammerman, 1984). Successive stages of degradation depend on the chemistry of the liquid phase and the bacterial species. For example, acetogenic bacteria can convert

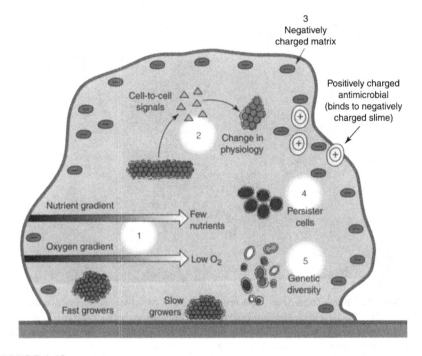

FIGURE 1-13   Biofilms derive their extraordinary tolerance to antimicrobial compounds from several factors. Bacteria near the center of a microcolony grow very slowly because they are exposed to lower concentrations of oxygen and nutrients (**1**). They are thus spared the effects of antibiotic drugs, which are much more effective against fast-growing cells. Intercellular signals (**2**) can alter the physiology of the biofilms, causing members to produce molecular pumps that expel antibiotics from the cells and allow the community to grow even in the presence of a drug. The biofilms matrix is negatively charged (**3**) and so binds to positively charged antimicrobials, preventing them from reaching the cells within the colony. Specialized populations of persister cells (**4**) do not grow in the presence of an antibiotic, but neither do they die. When the drug is removed, the persisters can give rise to a normal bacterial colony. This mechanism is believed to be responsible for recurrent infections in hospital settings. Finally, population diversity (**5**), genetic as well as physiological, acts as an "insurance policy," improving the chance for some cells to survive any challenge. (Harrison et al., 2005. Reprinted with permission of the authors.)

nonfermentable compounds into acetic acid and hydrogen. Other microorganisms can consume acetate and hydrogen.

It is extremely difficult to predict the impact of biofilms on degradation processes. Cells within biofilms may be viable but nonculturable (Oliver, 1993; del Mar Lleò et al., 2000), making it impossible to detect their presence with traditional culture techniques. Microorganisms within biofilms are capable of maintaining environments at biofilm–surface interfaces that are radically different from the bulk in terms of pH, dissolved oxygen, and other organic and inorganic species. In some cases, these interfacial conditions could not be maintained in the bulk medium at room temperature near atmospheric pressure. As a consequence, microorganisms within biofilms produce reactions that are not predicted by thermodynamic arguments based on the chemistry of the bulk medium.

REFERENCES

Armon R, Starosvetsky J, Dancygier M, Starosvetsky D (2001). Adsorption of *Flavobacterium breve* and *Pseudomonas fluorescens* P17 on different metals: electrochemical polarization effect. *Biofouling*, **17**: 289–301.

Azam F, Ammerman JW (1984). *Flows of Energy and Materials in Marine Ecosystems.* New York: Plenum Press, p. 345.

Azeredo J, Oliveira R (2000a). The role of exopolymers in the attachment of *Sphingomonas paucimobilis*. *Biofouling*, **16**(1): 59–67.

Azeredo J, Oliveira R (2000b). The role of exopolymers produced by *Sphingomonas paucimobilis* in biofilm formation and composition. *Biofouling*, **16**(1): 17–27.

Beyenal H, Lewandowski Z (2002). Internal and external mass transfer in biofilms grown at various flow velocities. *Biotechnol. Prog.*, **18**: 55–61.

Bott TR, Gunatillaka M (1983). *Fouling of Heat Exchanger Surfaces.* New York: United Engineering Trustees, p. 727.

Bradshaw DJ, Marsh PD, Watson GK, Allison C (1997). Oral anaerobes cannot survive oxygen stress without interacting with facultative/aerobic species as a microbial community. *Lett. Appl. Microbiol.*, **25**: 385–87.

Busscher HJ, Geertsema-Doornbusch GI, van der Mei HC (1997). Adhesion of silicone rubber of yeasts and bacteria isolated from voice prostheses: influence of salivary conditioning films. *J. Biomed. Mater. Res.*, **34**(2): 201–209.

Characklis WG Roe FL, Turakhia MH, Zelver N (1983). *Microbial Fouling and Its Effect on Power Generation.* Final Report, Office of Naval Research, NOOO14-80-C-0475.

Cole JJ (1982). Interactions between bacteria and algae in aquatic systems. *Annu. Rev. Ecol. Syst.*, **13**: 291–314.

Cowan MM, Warren TM, Fletcher M (1991). Mixed species colonization of solid surfaces in laboratory biofilms. *Biofouling*, **3**: 23–34.

del Mar Lleò M, Pierobon S, Tafi MC, Signoretto C, Canepari P (2000). mRNA detection by reverse transcription-PCR for monitoring viability over time in an *Enterococcus faecalis* viable but nonculturable population maintained in a laboratory microcosm. *Appl. Environ. Microbiol.*, **66**(10): 4564–4567.

Fleming HC (1996). Economical and technical overview. In: Heitz E, Fleming HC, Sand W (eds) *Microbially Influenced Corrosion of Materials.* New York: Springer-Verlag, pp. 6–14.

Fletcher M. (1988). Attachment of *Pseudomonas fluorescens* to glass and influence of electrolytes on bacterium-substratum separation distance. *J. Bacteriol.*, **170**: 2027–2030.

Flint SH, Bremer PJ, Brooks JD (1997). Biofilms in dairy manufacturing plant—description, current concerns and methods of control. *Biofouling*, **11**(1): 81–97.

Gerchakov SM, Roth FJ, Sallman B, Udey LR, Marszalek DS (1977). Observations on microfouling applicable to OTEC systems. In: Gray H (ed) *Proceedings of the Ocean Thermal Energy Conversion (OTEC) Biofouling and Corrosion Symposium*, Seattle, WA, pp. 63–75.

Gold J (1999). Micro- and nano- patterned biomaterials surfaces. *8th European Conference on Applications of Surfaces and Interface Analysis* (ECASIA99), p. 19.

Haack TK, McFeters GA (1982). Nutritional relationship among microorganisms in an epilithic biofilm community. *Microb. Ecol.*, **8**: 115.

Harrison JJ, Turner RJ, Marques LLR, Ceri H (2005). Biofilms. *Am. Sci.*, **93**(6): p. 513.

Kennedy JF, Barker SA, Humphreys JD (1976). The immobilization of microbial cells on some metal hydroxides. *Nature*, **261**: 242.

Korber DR, Chai A, Woolfaardt GM, Ingham SC, Cadwell DE (1997). Substratum topography influences susceptibility of *Salmonella enteritidis* biofilms to sodium phosphate. *Appl. Environ. Microbiol.*, **63**: 3352–3358.

Kulakov LA, McAlister MB, Ogden KL, Larkin MJ, O'Hanlon JF (2002). Analysis of bacteria contaminating ultrapure water in industrial systems. *Appl. Environ. Microbiol.*, **68**(4): 1548–1555.

Lewandowski Z (1998). Structure and function of bacterial biofilms, *Corrosion*, Paper No. 296, p. 15.

Lewandowski Z, Dirckx P (1996). *Biofilm Conceptual Illustration with Labels*. Center for Biofilm Engineering, Montana State University, Bozman, MT, viewed on June 1, 2006, <www.erc.montana.edu/Res-Lib99-SW/Image_Library/Structure-Function/Full-image%20pages/BiofilmWithLabelsjpg.htm>.

Lewandowski Z, Stoodley P (1995). Flow induced vibrations, drag force, and pressure drop in conduits covered with biofilm. *Water Sci. Technol.*, **32**(8): 19–26.

Little BJ, Wagner PA, Jacobus OJ (1988). The impact of sulfate-reducing bacteria on welded copper–nickel seawater piping systems. *Mater. Performance*, **27**(8): 57–61.

McEldowney S, Fletcher M (1988). Effects of pH, temperature and growth conditions on the adhesion of a gliding bacterium and three non gliding bacteria to polystyrene. *Microbiol. Ecol.*, **15**: 229.

McFeters GA, Broadway SC, Pyle BH, Siu KK, Egozy Y (1993). Bacterial ecology of operating laboratory water purification systems. *Ultrapure Water*, **10**: 32–37.

Medilanski E, Kaufmann K, Wick LY, Wanner O, Harms H (2002). Influence of the surface topography of stainless steel on bacterial adhesion, *Biofouling*, **18**(3): 193–203.

Nickels JS, Bobbie RJ, Martz RF, Smith GA, White DC, Richards NL (1981). Effects of silicate grain shape, structure, and location on the biomass and community structure of colonizing marine microbiota. *Appl. Environ. Microbiol.*, **41**:1262–1268.

Oliver JD (1993). Formation of viable but nonculturable cells. In: Kjelleberg S (ed.) *Starvation in Bacteria*. New York: Plenum Press, pp. 239–272.

Ostuni E, Chapman RG, Liang MN, Meluleni G, Pier G, Ingber DE, Whitesides GM (2001). Self-assembled monolayers that resist the adsorption of proteins and the adhesion of bacterial and mammalian cells. *Langmuir*, **17**(20): 6336–6343.

Poleunis C, Compere C, Bertrand P (2002). Time-of-flight secondary ion mass spectrometry: characterisation of stainless steel surfaces immersed in natural seawater. *J. Microbiol. Methods*, **48**(2–3): 195–205.

Pope DH (1986). *A Study of Microbiologically Influenced Corrosion in Nuclear Power Plants and a Practical Guide for Countermeasures*. Electric Power Research Institute Report NP-4582, Palo Alto, CA.

Sreekumari KR, Nandakumar K, Kikuchi Y (2001). Bacterial attachment to stainless steel welds: significance of substratum microstructure. *Biofouling*, **17**: 303–316.

Stoodley P, Dirckx P (2003a). *Biofilm Life Cycle in Three Steps*. Center for Biofilm Engineering, Montana State University, Bozman, MT, viewed on June 1, 2006, <www.erc.montana.edu/Res-Lib99-SW/Image_Library/Structure-Function/Full-image%20pages/CBE-03_BFin3steps.htm>.

Stoodley P, Dirckx P (2003b). *Biofilm Migrates in Various Ways*. Center for Biofilm Engineering, Montana State University, Bozman, MT, viewed on June 1, 2006, <www.erc.

montana.edu/Res-Lib99-SW/Image_Library/Structure-Function/Full-image%20pages/ CBE-03_BfMigration.htm>.

van Loosdrecht MC, Eikelboom MD, Gjaltema A, Mulder A, Tijhuis L, Heijnen JJ (1995). Biofilm structures. *Water Sci. Technol.*, **32**(8): 35–43.

Verran J, Boyd RD (2001). The relationship between substratum surface roughness and microbiological and organic soiling: a review. *Biofouling*, **17**(1): 59–71.

Wiencek K, Fletcher M (1997). Effects of substratum wettability and molecular topography on the initial adhesion of bacteria to chemically defined substrata, *Biofouling*, **11**(4): 293–311.

Wilkinson JF (1958). The extracellular polysaccharides of bacteria. *Bacteriol. Rev.*, **22**: 46–73.

Wimpenny J (1996). Ecological determinants of biofilm formation, *Biofouling*, **10**(1–3): 43–63.

Chapter 2

Causative Organisms and Possible Mechanisms

INTRODUCTION

The organisms known to influence corrosion are physiologically diverse and have frequently been grouped by either an electron acceptor or an energy source that is linked to the resulting corrosion, for example, sulfate-reducing, iron-oxidizing, and manganese-oxidizing bacteria (Table 2-1). However, it is recognized that while the corrosion may be attributed to a single group of organisms, the most aggressive microbiologically influenced corrosion (MIC) occurs with natural populations containing many types of microorganisms. Furthermore, a single type of microorganism can simultaneously affect corrosion via several mechanisms. Cell death or lysis within a well-developed biofilm does not necessarily mean a cessation of the influence on electrochemical processes. Miller and Tiller (1970) confirmed that pitting corrosion continues under deposits of iron-oxidizing bacteria, independent of biochemical activity of the bacteria. Similarly, microbiologically generated FeS accelerates corrosion reactions in the absence of viable cells (Booth and Tiller, 1962). The following is a discussion of causative organisms typically associated with MIC and mechanisms for MIC.

ENNOBLEMENT

No other phenomenon has fascinated those studying MIC more than ennoblement, that is, the increase of corrosion potential E_{corr} due to the formation of a biofilm on a metal surface (Figure 2-1). Microbial colonization of passive metals can shift E_{corr} in the noble direction and produce accompanying increases in current density and polarization slope at mild cathodic overpotentials. Ennoblement has been observed in fresh, estuarine, and marine waters with many metals and alloys. The phenomenon

Microbiologically Influenced Corrosion By Brenda J. Little and Jason S. Lee
Published 2007 by John Wiley & Sons, Inc.

TABLE 2-1   Common Microoganisms Identified with Microbiologically Influenced Corrosion

Genus or species	pH	Temperature (°F)	Oxygen requirement	Metals affected	Metabolic process
Bacteria					
Desulfovibrio	4–8	50–105	Anaerobic	Iron and steel, stainless steels, aluminum, zinc, copper alloys	Use hydrogen in reducing SO_4^{2-} to S^{2-} and H_2S, promote formation of sulfide films
Desulfoto-maculum	6–8	50–105 (some at 115–165)	Anaerobic	Iron and steel, stainless steels	Reduce SO_4^{2-} to S^{2-} and H_2S
Desulfomonas		50–105	Anaerobic	Iron and steel	Reduce SO_4^{2-} to S^{2-} and H_2S
Acidithio-bacillus thiooxidans	0.5–8	50–105	Aerobic	Iron and steel, copper alloys, concrete	Oxidize sulfur and sulfides to form H_2SO_4; damages protective coatings
Acidithio-bacillus ferrooxidans	1–7	50–105	Aerobic	Iron and steel	Oxidize Fe (II) to Fe (III)
Gallionella	7–10	70–105	Aerobic	Iron and steel, stainless steels	Oxidize Fe (II) and Mn (II) to Mn (IV); promotes tubercle formation
Siderocapsa	–	–	Micro-aerophilic	Iron and carbon steel	Oxidize iron
Leptothrix	6.5–9	50–95	Aerobic	Iron and steel	Oxidize Fe (II) and Mn (II) to Mn (IV)
Sphaerotilus	7–10	70–105	Aerobic	Iron and steel, stainless steels	Oxidize Fe (II) and Mn (II) to Mn (IV); promotes tubercle formation

Continued

TABLE 2-1 *Continued*

Genus or species	pH	Temperature (°F)	Oxygen requirement	Metals affected	Metabolic process
Sphaerotilus natans	–	–	–	Aluminum alloys	–
Pseudomonas	4–9	70–105	Aerobic	Iron and steel, stainless steels	Some strains reduce Fe^{3+} to Fe^{2+}
Pseudomonas aeruginosa	4–8	70–105	Aerobic	Aluminum alloys	–
Fungi					
Hormoconis resinae	3–7	50–115 (best at 85–95)	Aerobic	Aluminum alloys	Produce organic acids when metabolizing certain fuel constituents

Source: Adapted from Dexter (2003).

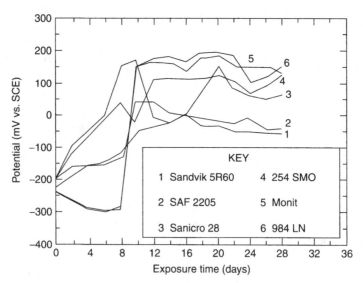

FIGURE 2-1 Open-circuit potential as function of time for six stainless steels exposed to flowing natural seawater Sandvik 5R60 (UNS S31603), SAF 2205 (UNS N08904), Sanicro 28 (UNS N08028), 254 SMO (UNS S31254), Monit (UNS S44635), NU984LN (non-UNS designation). (Johnsen and Bardal, 1985. © NACE International, 1985.)

is particularly important for alloys that have a pitting potential (E_{pit}) a few hundred millivolts more noble than E_{corr}, for example, 300 series stainless steels. The mechanism for ennoblement in fresh and estuarine waters is related to microbial manganese deposition and will be discussed in a later section under microbial metal oxidation.

Despite the extensive literature on the subject, the exact mechanism of ennoblement of metals in seawater remains unresolved. Ennoblement in marine waters has been attributed to organometallic catalysis, acidification of the electrode surface, the combined effects of elevated H_2O_2 and decreased pH, and the production of passivating siderophores. The susceptibility of stainless steels, aluminum, and nickel alloys to localized corrosion is often determined in laboratory tests by measurement of E_{pit} above which pits can initiate and grow. The dependence of E_{pit} for stainless steel on the activity of the chloride ion (a_{Cl^-}) is given by

$$E_{pit} = a - b \log a_{Cl^-} \qquad (2\text{-}1)$$

where a and b are experimentally determined parameters (Uhlig and Revie, 1985). The ennoblement of E_{corr} is usually explained by the acceleration of the cathodic oxygen reduction reaction due to microbial activities:

$$O_2 + 2H_2O + 4e^- \rightleftharpoons 4OH^- \qquad (2\text{-}2)$$

Either thermodynamic or kinetic effects can accelerate oxygen reduction and result in an ennoblement of E_{corr}. In the first case, localized acidification or an increase in the partial pressure of oxygen (p_{O_2}) increases the reversible potential of the oxygen electrode. Increased p_{O_2} leads to small changes in E_{corr}.

Acidification causes E_{pit} to become more negative. Mansfeld and Little (1991) showed that E_{pit} for 304 stainless steel (UNS S30400) decreased in deaerated 3.5% NaCl from about +300 mV versus the reference saturated calomel electrode (SCE) at pH 8 to about 0 mV at pH 2. On the basis of these results and the fact that E_{pit} has to be more positive than E_{corr}, it is unlikely that the ennoblement of E_{corr} is the result of microbial acid production.

CONCENTRATION CELLS

Oxygen Concentration Cells

Any geometrical factor that results in a high oxygen concentration in one area and a low concentration at another will create a differential cell. The physical presence of microbial cells on the surface, in addition to their metabolic activities, modifies electrochemical processes. Adsorbed cells grow, reproduce, and form colonies that are physical anomalies on a metal surface, resulting in local anodes and cathodes and the formation of differential aeration cells. Under aerobic conditions, areas under respiring colonies become anodic and surrounding areas become cathodic (Figure 2-2). A mature biofilm prevents the diffusion of oxygen to cathodic sites and the diffusion of aggressive anions, such as chloride, to anodic sites. Outward diffusion of metabolites and corrosion products is also impeded. If areas within the biofilm become anaerobic, that is, if the aerobic respiration rate within the biofilm is greater than the

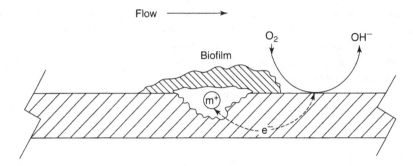

FIGURE 2-2   The physical presence of microbial cells on a metal surface, as well as their metabolic activities, modifies electrochemical processes. The adsorbed cells grow and reproduce, forming colonies that constitute physical anomalies on a metal surface, resulting in the formation of local cathodes or anodes.

oxygen diffusion rate, a change in the cathodic mechanism occurs (Characklis and Marshall, 1990).

Metal Concentration Cells

All microorganisms that colonize metal surfaces produce polymers and form a gel matrix on the metal. Microorganisms that produce copious amounts of extracellular polymeric substances (EPS) are generically referred to as "slime-forming" bacteria. In general, EPS are acidic and contain functional groups that bind metals (Geesey et al., 1986). Metal ions concentrated from the aqueous phase or the substratum into the biofilm increase corrosion rates by providing an additional cathodic reaction. Geesey and Bremer (1999) used Fourier transform infrared spectroscopy to characterize binding of an acidic polysaccharide to thin films of copper. Their measurements suggest a copper ion interaction with carboxyl groups of the polymer that promotes the ionization of the metallic copper (Figure 2-3). Ford et al. (1990) investigated the relationship between EPS and corroding metal surfaces. Their studies established that there are considerable differences in binding capacity between individual metal ions and specific exopolymers. Nivens et al. (1986) demonstrated that *Vibrio natriegens* increased the corrosion rate of S30400 stainless-steel during a 6-day incubation. The corrosion rate began to increase when colonies of microorganisms were detected on the surface. The most rapid increase in corrosion rate, however, correlated with the formation of EPS.

REACTIONS WITHIN BIOFILMS

Reactions within biofilms are generally localized, affecting mechanisms and accelerating rates of electrochemical reactions leading to corrosion.

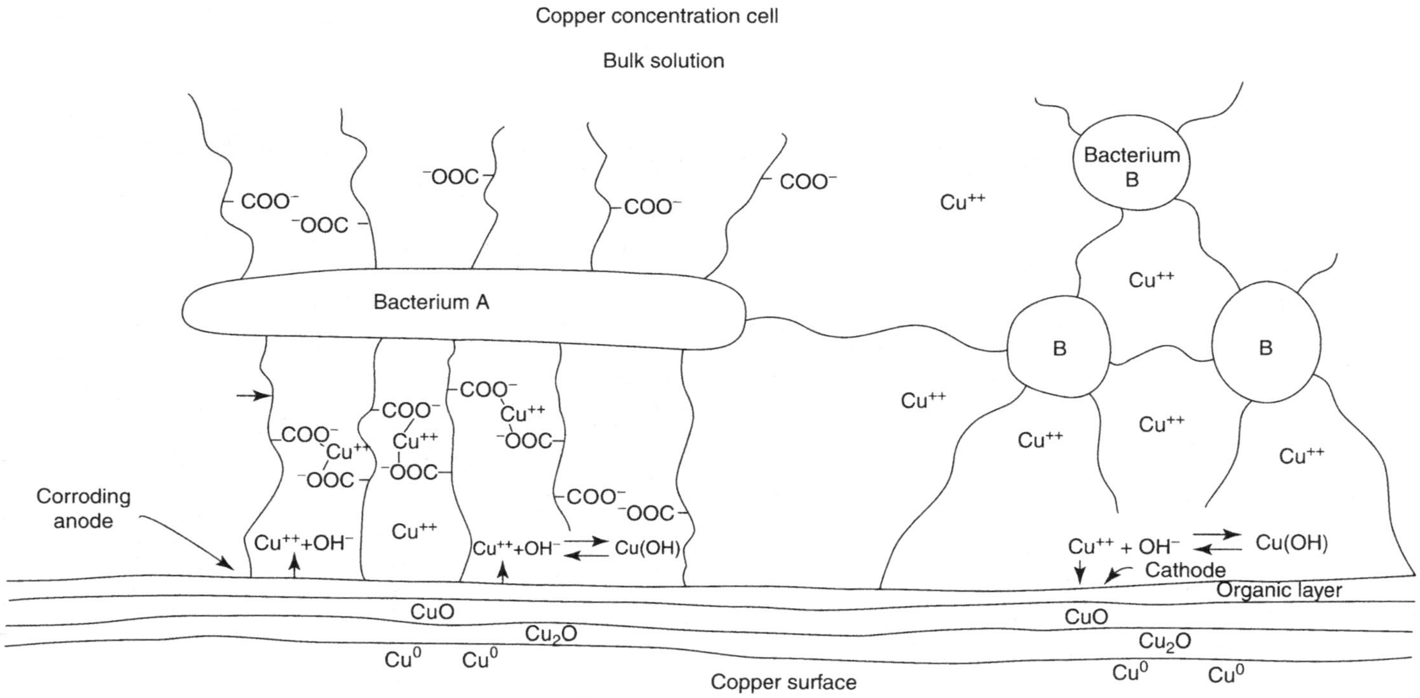

FIGURE 2-3 Copper concentration cell. (Geesey et al., 1986. © NACE International, 1986.)

27

Respiration/Photosynthesis

Algae and photosynthetic bacteria use light to produce oxygen that can accumulate within a biofilm. Over the normal range of pH values encountered in seawater, the availability of oxygen has a greater influence on corrosion than pH (Laque, 1975). Increased oxygen concentration can depolarize the cathodic reaction, leading to increased corrosion rates. During dark periods algae respire, converting oxygen into CO_2. Localized respiration/photosynthesis can lead to differential aeration cells and localized anodes and cathodes.

Sulfide Production

Reduction of elemental sulfur or thiosulfate ($S_2O_3^{2-}$) results in the production of hydrogen sulfide (H_2S), which acidifies a corrosive medium and catalyzes the penetration of hydrogen into steels, a process known as H_2S-induced cracking or sulfide-induced stress cracking. Crolet and Magot (1995) described a group of bacteria isolated from an oilfield production facility capable of reducing thiosulfate, not sulfate (SO_4^{2-}), to sulfide (S^{2-}).

Sulfate-reducing bacteria (SRB) are the organisms most closely identified with MIC. They are a group of ubiquitous, diverse anaerobes that use the sulfate ion as the terminal electron acceptor, producing H_2S. Several SRB can reduce nitrate, sulfite, thiosulfate, or fumarate with organic compounds or H_2. Sulfate-reducing bacteria have been isolated from a variety of environments (Pfennig et al., 1981; Postgate, 1979) including seawater where the concentration of sulfate is typically 25 mM (Postgate, 1979). Even though the oxygen content of seawater above the thermocline ranges from 5 to 8 ppm, anaerobic microorganisms survive in anaerobic microniches until conditions are suitable for their growth (Costerton and Geesey, 1986; Staffeldt and Kohler, 1973). If the aerobic respiration rate within a biofilm is greater than the oxygen diffusion rate, the metal–biofilm interface can become anaerobic and provide a niche for sulfide production by SRB (Figure 2-4) (Hamilton, 1985). The critical biofilm thickness required to produce anaerobic conditions depends on availability of oxygen and the respiration rates of organisms in the biofilm. The metabolic activity of SRB causes accumulation of sulfide near metal surfaces. Sulfate-reducing bacteria have been the focus of many investigations involving MIC, and several corrosion mechanisms have been attributed to SRB, including cathodic depolarization by the enzyme dehydrogenase, anodic depolarization, production of iron sulfides, release of exopolymers capable of binding metal ions, sulfide-induced stress corrosion cracking, and hydrogen-induced cracking or blistering. Recent reviews suggest that SRB can influence a number of corrosion mechanisms simultaneously.

The early work of von Wolzogen Kuhr and van der Vlugt (1934) with SRB is often cited as the initial demonstration of MIC. They suggested the following electrochemical reactions:

$$4Fe \rightarrow 4Fe^{2+} + 8e^- \text{ (anodic reaction)} \tag{2-3}$$

$$8H_2O \rightleftharpoons 8H^+ + 8OH^- \text{ (water dissociation)} \tag{2-4}$$

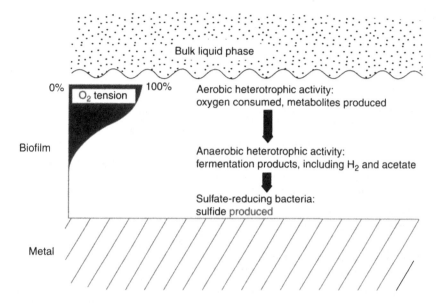

FIGURE 2-4 A schematic diagram of the spatial relationship between aerobes, heterotrophic anaerobes, and sulfate-reducing bacteria in a biofilm accumulated on a metal surface. (Reprinted from Hamilton, 1985, with permission from the *Annual Review of Microbiology*; www.annualreviews.org.)

$$8H^+ + 8e^- \rightleftharpoons 8H \text{ (ads) (cathodic reaction)} \qquad (2\text{-}5)$$

$$SO_4^{2-} + 8H \rightleftharpoons S^{2-} + 4H_2O \text{ (bacterial consumption)} \qquad (2\text{-}6)$$

$$Fe^{2+} + S^{2-} \rightleftharpoons FeS \text{ (corrosion products)} \qquad (2\text{-}7)$$

$$4Fe + SO_4^{2-} + 4H_2O_3 \rightleftharpoons Fe(OH)_2 + FeS + 2OH^- \qquad (2\text{-}8)$$

They described the overall process as cathodic depolarization, based on the theory that hydrogenase-positive (Hase$^+$) SRB remove hydrogen accumulated at the cathode. Removal of cathodic hydrogen in step (2-5) forces iron to dissolve at the anode in step (2-3). While the overall reaction may be described correctly in step (2-8), it is doubtful that individual reaction steps (2-3) to (2-7) proceed in the manner proposed by von Wolzogen Kuhr and van der Vlugt (1934). It is unlikely that a layer of atomic hydrogen exists on the metal surface as postulated in step (2-5) (Schweisfurth and Heitz, 1989). The concentration of H$^+$ is extremely small in anaerobic, neutral environments. Therefore, additional cathodic reactions should be considered such as the reduction of H_2S:

$$H_2S + e^- \rightleftharpoons HS^- + \frac{1}{2} H_2 \qquad (2\text{-}9)$$

Hamilton (1985) discussed H_2S reduction as the cathodic reaction and the role of FeS as a cathode with a low hydrogen overvoltage. Furthermore, strains of the SRB *Desulfovibrio* that do not produce hydrogenase (Hase⁻) can stimulate corrosion by cathodic depolarization induced by microbiologically produced FeS. King et al. (1973) demonstrated that weight loss of steel was proportional to the concentration of ferrous sulfide present and depended on the stoichiometry of the particular ferrous sulfide minerals.

Most of the laboratory work with SRB has concentrated on the *Desulfovibrio* species. However, Dinh (2004) found that *Desulfobacterium* dominated a natural marine population and grew faster than the *Desulfovibrio* species. McNeil and Odom (1994) developed a thermodynamic model for predicting SRB-influenced corrosion based on the likelihood that a metal would react with microbiologically produced sulfide. The model is based on the assumption that SRB MIC is initiated by sulfide-rich reducing conditions in the biofilm and that under those conditions, the oxide layer on the metal (or the metal itself) is destabilized and acts as a source of metal ions. At the outer surface of the SRB, these ions react to produce sulfide compounds in micron-sized particles that are in some cases crystalline (Figure 2-5). The consumption of metal ions at the microbe surface is balanced by release of surface ions until the oxide is totally consumed. If the reaction to convert the metal

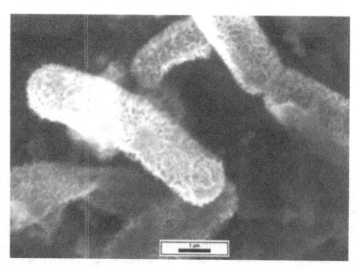

FIGURE 2-5   Crystalline deposits on SRB at the surface of 90/10 copper-nickel alloy (UNS C70600) exposed to natural seawater. (Little et al., 1991.)

oxide into a metal sulfide has a positive Gibbs free energy under surface conditions, the sulfides will not strip the protective oxide and no corrosion will take place. If the Gibbs free energy is negative, this reaction will proceed, and the sulfide microcrystal will redissolve and reprecipitate as larger, generally more sulfur-rich crystals, ultimately altering the sulfide minerals that are stable under biofilm conditions. In the presence of dissolved oxygen, the sulfides may react to form other compounds. The impact of oxygen on corrosion by SRB will be discussed in detail later in this chapter. McNeil and Odom prepared the following database based on their model:

- *Ag*   Acanthite (Ag_2S).
- *Ag–Cu alloys*   Acanthite, argentite (the high-temperature polymorph of Ag_2S or jalpaite – Ag_3CuS_2).
- *Cu*   Complex suites of sulfide minerals: The most common product is chalcocite(Cu_2S). Final product in many cases is blue-remaining covellite (CuS_{1+x}).
- *Cu–Ni alloys*   Sulfide corrosion products similar to those of Cu but with significant djurleite ($Cu_{31}S_{16}$). No Ni minerals observed.
- *Cu–Sn alloys*   Corrosion products similar to those in Cu.
- *Fe (carbon steel)*   Final product is pyrite (FeS) with numerous intermediates.
- *Fe (stainless alloys)*   Rates are slower than pure Fe or carbon steel. No Ni minerals have been detected. Stainless steels with 6 percent or more Mo appear to be very resistant.
- *Ni*   Millerite (NiS).
- *Pb*   Galena (PbS).

The model predicts that titanium alloys will be immune to reactions of sulfide along with most stainless steels. The model is limited to thermodynamic predictions as to whether a reaction will take place and does not consider metal toxicity to the organisms, tenacity of the resulting sulfide, or other factors that influence corrosion rate.

Many sulfide minerals under near-surface natural environmental conditions can only be produced by microbiological action on specific precursor materials such as metals. Consequently, sulfides can provide a mineralogical fingerprint for SRB-induced corrosion. In the following sections, SRB sulfide production will be reviewed for iron, copper, copper alloys, silver, zinc, and lead. The metal interface under the biofilm and corrosion layers will be referred to as base metal to differentiate it from layers of minerals and metal ions that have been derivatized by corrosion reactions. Mineralogical data, thermodynamic stability diagrams (Pourbaix, 1966; Wagman et al., 1982), and the simplexity principle for precipitation reactions (McNeil et al., 1991) will be used to rationalize corrosion product mineralogy to demonstrate the action of SRB.

Iron

The corrosion rate of iron in the presence of H_2S is accelerated by the formation of iron sulfide minerals (Wikjord et al., 1980). Once electrical contact is established, carbon steel behaves as an anode and electron transfer occurs through the iron sulfide. In the absence of oxygen, the metabolic activity of SRB causes accumulation of H_2S near metal surfaces. This is particularly evident when metal surfaces are covered with biofilms. Figure 2-6 shows concentration profiles of sulfide, oxygen, and pH in a biofilm accumulated on the surface of a carbon steel (UNS G10180) corrosion coupon. The concentration of sulfide is highest near the metal surface where iron sulfide forms quickly and covers the steel surface if both ferrous and sulfide ions are available. At low ferrous ion concentrations, adherent and temporarily protective films of iron sulfides are formed on the steel surface with a consequent reduction in corrosion rate.

Figure 2-7 is a stability diagram for an iron–water–reduced sulfur system with lines for 10^{-6} M ferrous iron and 10^{-2} M sulfide. For clarity, pyrite (FeS_2) and mackinawite [$FeS_{(1-x)}$] are the only sulfides indicated. Parallelograms superimposed on the diagram are bounded by the highest and lowest pH values commonly found in natural fresh and saline surface waters. The upper portion of the hatched area applies to waters less than 10 m from the surface; the lower portion (oppositely hatched) represents waters at depths greater than 10 m. Conditions in the upper-hatched parallelogram represent those readily achieved in stagnant waters. The lower portion indicates conditions not found in near-surface environments. Mackinawite is a

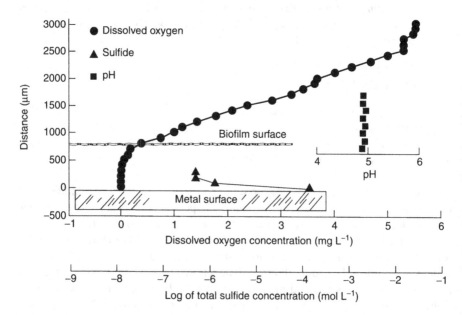

FIGURE 2-6   Concentration profiles of sulfide, oxygen, and pH in a biofilm on mild steel (UNS G1018). (Reprinted from Lee et al., 1993, with permission from Taylor and Francis Ltd. http://www.tandf.uk/journals.)

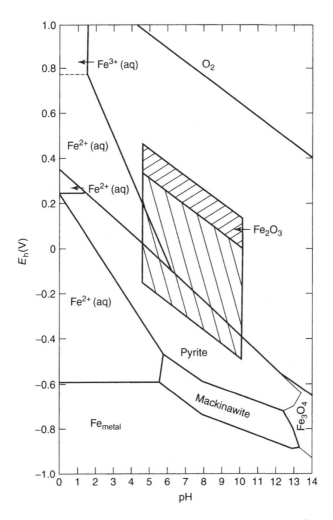

FIGURE 2-7   Iron stability diagram: water without chloride ions; total sulfide 10^{-2} M. (McNeil et al., 1991. © NACE International, 1991.)

tetragonal mineral that cannot be produced by conventional techniques. Greigite is a thiospinel with formula Fe_3S_4. Smythite is a hexagonal compound with formula $Fe_{3-x}S_4$, $0 < x < 0.25$. Cubic FeS can be produced artificially. Thermodynamic analyses indicate that under redox and sulfide activity conditions in surface waters, only pyrite is stable; furthermore, pyrite forms relatively easily in nonbiological corrosion, so the preferential formation of less stable sulfides is difficult to attribute to slow pyrite formation kinetics. The region of stability for pyrite is accessible under severe conditions. The region of stability of mackinawite is wholly outside the region defined by surface water conditions, excluding waters influenced by peat bogs, coal mines, volcanic activity, and industrial effluents.

During corrosion of iron and steel in the presence of SRB, a thin (~1 μm) adherent layer of mackinawite is formed. As it thickens, the layer becomes less adherent. If ferrous ion concentration in the electrolyte is low, mackinawite alters to greigite. This alteration is not observed in nonbiological systems. If the ferrous ion concentration is high, mackinawite is accompanied by green rust 2, a complex ferrosoferric oxyhydroxide.

The impact of oxygen on corrosion resulting from obligate anaerobic SRB was examined by Hardy and Bown (1984) using carbon steel (undesignated) and weight-loss measurements. Successive aeration–deaeration shifts caused variation in the corrosion rate. The highest corrosion rates were observed during periods of aeration. Lee et al. (1993) determined that corrosion of carbon steel could not be initiated by SRB in the absence of ferrous ion. In their experiments, there was no cathodic depolarization and no correlation between corrosion rates and SRB in the absence of ferrous ion. At low ferrous ion concentrations (0–10 mg L^{-1}), adherent and temporarily protective films of iron sulfides form on the steel surface with a consequent reduction in anodic and cathodic currents. High corrosion current densities associated with SRB-induced corrosion of carbon steel were maintained only in high concentrations of ferrous ion.

In summary, mackinawite (tetragonal FeS_{1-x}) is easily produced from iron and iron oxides by consortia of microorganisms that include SRB. The presence of mackinawite in corrosion products formed in shallow water environments with the exclusions previously delineated is proof that the corrosion was SRB-induced. Recent work indicates that on continued exposure to SRB, mackinawite alters to greigite (Fe_3S_4), smythite (Fe_9S_{11}), and finally to pyrrhotite (FeS_{1+x}) (McNeil and Little, 1990). SRB in thin biofilms on pottery surfaces (Duncan and Ganiaris, 1987; Heimann, 1989) and silver (McNeil and Mohr, 1993) can produce pyrite films from iron-rich waters. Pyrite is not a typical iron corrosion product, but SRB can produce pyrite from mackinawite in contact with elemental sulfur (Berner, 1969). Abiotic aqueous synthesis of these minerals, with the possible exception of pyrite, requires H_2S pressures higher than those found in shallow waters.

Copper

Cuprite (Cu_2O), the first product of copper corrosion, forms epitaxially as a direct reaction product of copper with dissolved O_2 or water molecules (North and Pryor, 1970). Cuprite has a high electrical conductivity and permits transport of copper ions through the oxide layer so that they can dissolve in the water and reprecipitate. If the water chemistry approximates that of seawater, copper ions reprecipitate as botallackite ($Cu_2(OH)Cl$) (Pollard et al., 1989) that can alter in minutes or hours to either paratacamite or atacamite (other crystal structures of $Cu_2(OH)_3Cl$), depending on local water chemistry.

Figure 2-8 is a stability diagram for copper and its minerals drawn for 10^{-6} M total dissolved copper and 10^{-2} M total sulfide. Parallelograms superimposed on the diagram are similar to those described for Figure 2-7 and are appropriate for the analysis of corrosion mineralogy under nonhydrothermal conditions.

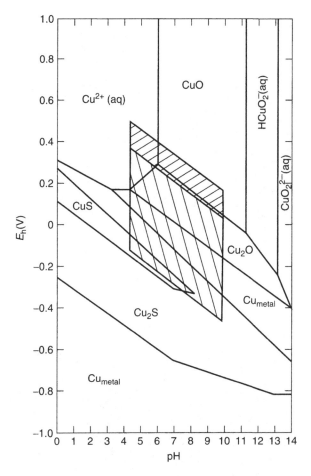

FIGURE 2-8   Copper stability diagram: water without chloride ions; total sulfide 10^{-2} M. (McNeil et al., 1991. © NACE International, 1991.)

McNeil et al. (1991) used Figure 2-8 to interpret results from laboratory experiments. They exposed mixed cultures known to contain SRB to copper (UNS C11000) and copper-nickel alloys 90/10 (UNS C70600) and 70/30 (UNS C71500) in a variety of natural and synthetic waters containing sulfates for 150 days. The pH values of the waters, measured after 2 weeks, were between 5.5 and 6.8. All copper-containing metals exposed to SRB in isolated cultures and in the natural augmented waters were covered with black sulfur-rich deposits (Figure 2-9). The thickness and tenacity of the surface deposits varied among the metals and cultures. Corrosion products on commercially pure copper (C11000) were consistently nonadherent. Corrosion products on copper alloys were more adherent and in some cases difficult to scrape from the surface. In all cases, bacteria were closely associated with sulfur-rich deposits (Figure 2-10), with many bacteria

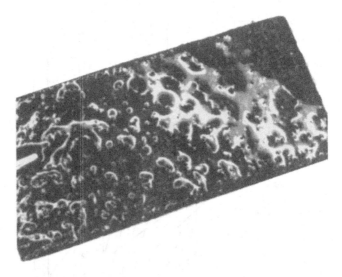

FIGURE 2-9   Black sulfide film on UNS C11000 copper foil after 4-month exposure to SRB. (McNeil and Little, 1998.)

FIGURE 2-10   ESEM micrographs of bacteria encrusted with corrosion products on UNS C11000 copper foils exposed to natural seawater. (Little et al., 1991.)

encrusted with deposits of copper sulfides. Most environmental scanning electron microscopy (ESEM) micrographs of SRB on copper-containing surfaces indicate a monolayer of cells overlaying a sulfide layer. Transmission electron microscopy has been used to demonstrate that bacteria are intimately associated with sulfide minerals (Blunn, 1986) and is illustrated in Figure 2-11 (Chiou, 1996).

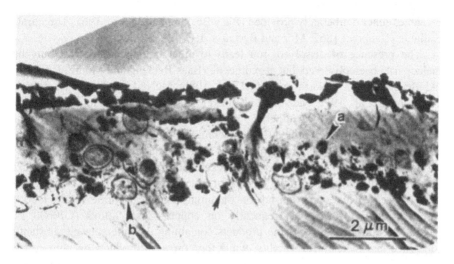

FIGURE 2-11   TEM micrograph showing bacteria on the surface and surrounded by a corrosion layer on 90/10 copper-nickel (UNS C70600) after 15 weeks exposure in seawater. (Chiou, 1996.)

Biomineralogy of copper sulfides has been studied for over a century (Daubree, 1862; de Gouvernain, 1875; Baas-Becking and Moore, 1961; Mor and Beccaria, 1975; Syrett, 1977, 1980, 1981; McNeil and Little, 1992). The complexity of the resulting observations reflects the complexity of the copper–sulfur system, especially in the presence of alloying elements or iron in the environment (Ribbe, 1976; Kostov and Minceva-Stefanova, 1981). Both high- and low-temperature polytypes of chalcocite (Cu_2S), digenite (Cu_9S_5), djurleite ($Cu_{193}S$-$Cu_{197}S$), anilite (Cu_7S_4), spionkopite ($Cu_{39}S_{28}$), geerite (Cu_8S_5), and covellite (CuS, generally blue-remaining) have been reported. In long-term corrosion where waters contain significant iron, chalcopyrite is a common product (Daubree, 1862; de Gouvernain, 1875; McNeil and Mohr, 1993). While chalcopyrite films can be formed abiotically in high sulfur concentrations (Cuthbert, 1962), chalcopyrite and most other copper sulfides are not generally found as products of abiotic corrosion.

Detailed kinetics of individual reactions are not fully understood, and the consequences of corrosion depend on many factors, including mineral morphology and variations of redox and pH with time (McNeil and Mohr, 1993). Discussions of alteration kinetics are contained in a number of papers (Baas-Becking and Moore, 1961; Roseboom, 1966; Craig and Scott, 1974; Putnis, 1977; Evans, 1979). The general phenomenology can be understood by the following approach: Microbial consortia that include SRB produce anoxic, sulfide-rich environments in which the conversion of copper into copper sulfides is thermodynamically favored at a concentration of 10^{-2} M total sulfur. Reactions appear to proceed as suggested by Ostwald's rule: The first sulfur-poor compounds are converted into sulfur-rich compounds. In short-term experiments with excess of copper over available sulfur, chalcocite with little or no covellite is formed (McNeil et al., 1991). Covellite is produced if excess sulfide is available, either deliberately provided (Baas-Becking and Moore, 1961) or naturally available (Daubree, 1862; Mor and Beccaria, 1975).

The presence of dissolved iron leads to other complications. Not only has chalcopyrite been observed, but also digenite (Baas-Becking and Moore, 1961; Mor and Beccaria, 1975; North and MacLeod, 1986; McNeil et al., 1991), djurleite (Macdonald et al., 1979; McNeil et al., 1991), and the hexagonal high-temperature polytype of chalcocite (McNeil et al., 1991). These observations can be interpreted in terms of the simplexity principle (Goldschmidt, 1953): Impurities tend to stabilize high-entropy, high-temperature polytypes. Digenite stability is promoted by iron (Craig and Scott, 1974). It appears that nickel stabilizes djurleite on copper–nickel and the stabilization of djurleite has major practical consequences. McNeil et al. (1991) observed that corrosion layers containing digenite were never observed on pure copper (C11000), but frequently on copper–nickel alloys (C70600 and C71500). Furthermore, corrosion products containing digenite showed substantial adherence and mechanical stability, while the corrosion products on pure copper, composed of chalcocite with only traces of other minerals, were powdery and non-adherent.

It has been argued that if the copper sulfide layer were djurelite, the sulfide layer would be protective (Nilsson et al., 1980). Even if such a sulfide film were technically passivating, the mechanical stability of the film is so poor that sulfide films are useless for corrosion protection. In the presence of turbulence, the loosely adherent sulfide film is removed, exposing a fresh copper surface to react with sulfide ions. For these reasons, turbulence-induced corrosion and sulfide attack of copper alloys cannot easily be decoupled. In the presence of oxygen, the possible corrosion reactions in a copper sulfide system are extremely complex because of the large number of stable copper sulfides (Ribbe, 1974), their differing electrical conductivities, and catalytic effects. Transformations between sulfides or of sulfides to oxides result in changes in volume that weaken the attachment scale and oxide subscale, leading to spalling. Bared areas repassivate, forming cuprous oxide (Figure 2-12a, b).

Silver

Figure 2-13 is a silver–water–chloride–sulfur system stability diagram. The upper diagonal line (a) is the oxygen line, above which water is thermodynamically unstable with respect to oxygen generation. The lower diagonal line (b) is the hydrogen line, below which water is thermodynamically unstable with regard to hydrogen evolution. Waters do exist outside these boundaries under special conditions. Horizontal line (c) separates the regions in which chloride corrosion of silver can and cannot take place. The upper-hatched area below line (d) approximates the region of effective redox–acidity conditions for near-surface waters. Strongly acidic and basic regions are characteristic of groundwaters, but not seawaters. The effective redox potential of oxygenated water is less than what would be calculated thermodynamically from oxygen concentrations because of kinetic effects. The effective redox potential also depends on water pollutants, temperature, and temperature–pressure

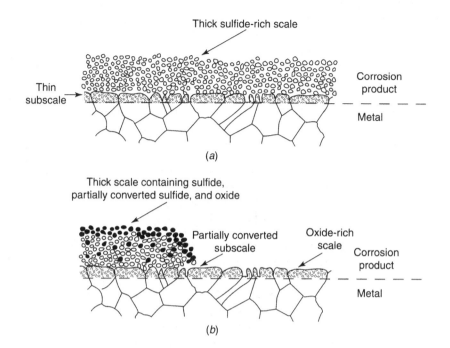

FIGURE 2-12 *a, b* Schematic of (*a*) thick sulfide-rich scale on copper alloy and (*b*) disruption of sulfide film. (Reprinted from Syrett, 1981, with permission from Elsevier.)

history. The lower, oppositely hatched parallelogram indicates redox–acidity conditions existing in natural waters but not characteristic of near-surface waters. Bulk water conditions outside the parallelograms are found in peat bogs, coal mines, and at depths not considered in this chapter. The heavy diagonal line (e) through the hatched area indicates redox conditions for a series of waters sampled from a brackish, stagnant pond in an industrialized area in Scandinavia where surface waters had a high oxygen content and deep waters contained 40 mg L^{-1} H$_2$S and significant amounts of decaying organic material (Garrels and Christ, 1965). Pond conditions were used to define the upper and lower redox potential (E_h) limits near pH 7. Upper and lower E_h limits at other pH values were estimated by assuming them to be controlled by oxidation. The hatched regions were generated using pond conditions to define the upper and lower E_h/pH values near neutrality. The upper portion of the hatched area applies for waters less than 10 m from the surface; the lower portion represents waters at depths greater than 10 m. Achieving conditions outside those defined by the two parallelograms at a metal surface requires the presence of a biofilm and maintenance of conditions radically different from those in the bulk environment.

Silver and its alloys are subject to corrosion by reduced sulfur species of microbiological origin. In air (e.g., in a museum) H$_2$S can be the consequence of biodegradation of sulfur-containing polymeric materials, producing monoclinic acanthite (Ag$_2$S) (Banister, 1952; Bauer 1988). Figure 2-13 indicates the possible thermodynamically stable phases for silver equilibrated with varying total sulfur compositions

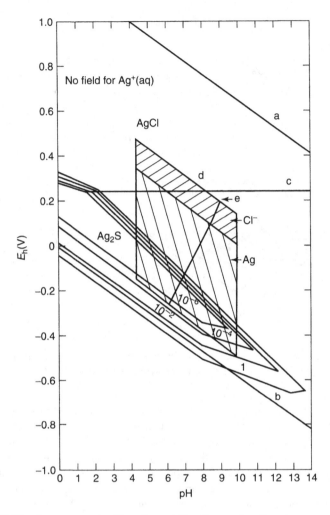

FIGURE 2-13   Silver–water–chloride–sulfur stability diagram for silver in seawater at varying reduced sulfur concentrations. (McNeil et al., 1991.)

in 0.46 M NaCl (typical for seawater). It is assumed that reduced sulfur species (S^{2-}, HS^-, H_2S) are in equilibrium. A straightforward type of silver corrosion is conversion of silver into cerargyrite (AgCl), as indicated above line (c). Below line (c) metallic silver is stable, except for the wedge-shaped areas pointing down and to the right indicating regions of stability for monoclinic acanthite (Ag_2S). The region between the diagonal lines bounding the upper-hatched region approximates the effective oxidizing behavior of near-surface, fully aerated seawater (Garrels and Christ, 1965). Most shallow sea chemistries fall into this region. Conditions in shallow land burials where the major source of groundwater is rain or surface water percolating through soils are near this region. Cerargyrite is stable in seawater and chloride-rich shallow-land burial conditions.

Other Metals

Sulfate-reducing bacteria-induced corrosion of zinc produces a zinc sulfide reported to be sphalerite (ZnS) (Baas-Becking and Moore, 1961). Sulfate-reducing bacteria on lead carbonates produces galena (PbS) (McNeil and Little, 1990). Galena has been found more recently as a lead corrosion product in SRB-induced corrosion of lead–tin alloys (McNeil and Mohr, 1993).

Acid Production

Elemental sulfur, thiosulfates, metal sulfides, H_2S, and tetrathionates can be oxidized to sulfuric acid by microorganisms generically referred to as thiobacilli or sulfur-oxidizing bacteria (SOB). Heterotrophic bacteria that secrete organic acids during fermentation of organic substrates are referred to as acid-producing bacteria (APB). The kinds and amounts of acids produced depend on the type of microorganisms and the available substrate molecules. Organic acids may force a shift in the tendency for corrosion to occur. The impact of acidic metabolites is intensified when they are trapped at the biofilm–metal interface. Acetic acid from *Clostridium aceticum* and sulfuric acid produced by SOB, such as *Thiobacillus thioxidans*, are obvious contributors to corrosion. In addition, it has been demonstrated that the organic acids of the Krebs cycle can promote the electrochemical oxidation of a variety of metals by removing or preventing the formation of an oxide film. Burns et al. (1969) showed that under aerobic conditions solutions of citric, fumaric, ketoglutaric, glutaric, maleic, malic, itaconic, pyruvic, and succinic acids formed metallic salts when incubated with copper, tin, and zinc. Little et al. (1986) demonstrated that isobutyric and isovaleric acids accelerate the corrosion of nickel. Gerchakov and Udey (1984) suggested that amino and dicarboxylic acids were also aggressive. Little et al. (1988) demonstrated that an aerobic acetic acid-producing bacterium accelerated corrosion of cathodically protected stainless steel (UNS N08366). Acetic acid destabilized or dissolved calcareous deposits formed during cathodic polarization and attacked the stainless-steel surface. Organic acids produced by fungi were identified as the cause for pitting failures in painted carbon steel (UNS G10200) holds of a bulk carrier (Stranger-Johannesen, 1986) and aluminum alloy 2024 (UNS A92024) fuel storage tanks (Salvarezza et al., 1983).

The pH under algal fouling varies with photosynthesis and respiration. Daily pH changes of up to 2 units have been recorded under algal cultures with pH values above 10 during photosynthesis (Terry and Edyvean, 1986).

Ammonia Production

Many organisms including nitrate-reducing bacteria can produce NH_3 from the metabolism of amino acids or the reduction of nitrite or nitrate forming NH_4^+. Nitrate-based corrosion inhibitors can be a source of nitrogen for microbial ammonia production. Pope et al. (1984a,b) generally discussed potential corrosion in copper alloys, particularly stress corrosion cracking, due to the activities of ammonia-producing bacteria.

Metal Deposition

Biomineralization of iron and manganese oxides occurs widely in natural waters, and is a dominant control in geochemical cycling of these elements (Gounot, 1994). Biomineralization can be carried out by a variety of organisms including bacteria, yeast, and fungi (Nealson et al., 1988). Ghiorse (1984) prepared a review of metal-depositing microorganisms in which he identified microorganisms that catalyze the oxidation of metals, others that accumulate abiotically oxidized metal precipitates, and still others that derive energy by oxidizing metals.

Manganese

Manganese is the third most abundant transition metal. Manganese oxidation is coupled to cell growth and metabolism of heterotrophic substrates (Arnold et al., 1988; Jung and Schweisfurth, 1976a,b). While the reduced form of manganese, Mn^{2+} (manganous), is soluble, all the various manganic oxidized forms, Mn_2O_3, $MnOOH$, Mn_3O_4, and MnO_2, are insoluble. Figure 2-14 shows regions of the United States where shallow groundwater manganese levels are significant. These regions include

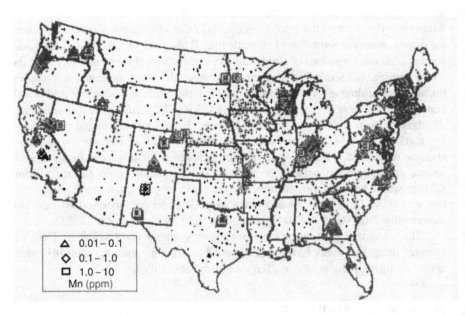

FIGURE 2-14   Comprehensive locator map for U.S. electric power stations (+). The manganese overlay map highlights regions with elevated manganese in shallow groundwater wells. The overlay is based on available data for selected aquifers and does not provide comprehensive geographic coverage. *Sources*: United States Department of Energy – Energy Information Administration, Form EIA-860A, Annual Electric Generator Report – Utility, 1999 and United States Department of Interior, U.S. Geological Survey – National Water Quality Assessment Program. (Dickinson et al., 2002. © NACE International, 2002.)

New England, the Connecticut and Ohio River valleys, the Delaware, Hudson, and Chesapeake River drainages, Western Georgia, Colorado, the Pacific Northwest, the Ozark Mountains, the Sierra Nevadas, Wisconsin, Texas, and New Mexico.

Microbially deposited manganese oxides have an amorphous structure as MnO_2 (vernadite) and sometimes form a black precipitate of MnO_2 (birnessite) found with *Leptothrix* and spores of *Bacillus* spp. (Gounot, 1994). In the *Bacillus*, birnessite recrystallizes to octahedral Mn_3O_4 (hausmannite) (Nealson et al., 1988). The relationship of manganese-depositing bacteria with MnO_2 can be demonstrated with x-ray diffraction. In Figure 2-15, a *Pseudomonas*-like organism, originally isolated from freshwater (Jung and Schweisfurth, 1976a,b), is imbedded in the manganese oxides it produced.

As a result of microbial action, manganese oxide deposits are formed on submerged materials including metal, stone, glass, and plastic and can occur in natural waters with manganese levels as low as 10 to 20 ppb (Dickinson and Lewandowski, 1996). Deposition rates of 1 mC cm^{-2} day^{-1} on stainless steel (UNS S31603) have been observed (Dickinson et al., 1996). Mature manganese deposits from both freshwater and marine sources can be identified as ring structures that become more and more numerous and dense with time (Figure 2-16a, b). In some instances, surfaces become covered with porous, blackened microbiologically mediated manganese dioxide deposits (Figure 2-17).

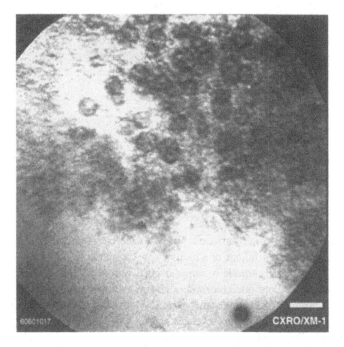

FIGURE 2-15   X-ray micrograph of manganese depositing bacteria with Mn-oxides. Cells are dark ellipse, oxides are lighter material. (Little et al., 1998.)

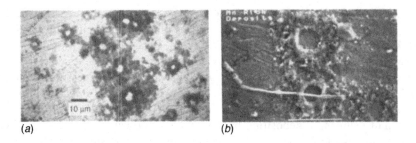

(a)   (b)

FIGURE 2-16a, b Biodeposited manganese oxide on 316L stainless steel (UNS S31603) exposed to fresh water. (Reprinted from Dickinson et al., 1996, with permission from Elsevier.)

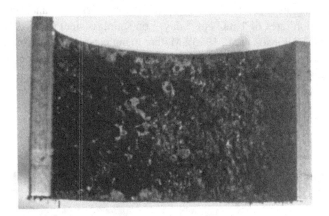

FIGURE 2-17 Black manganese deposits. (Little et al., 1998.)

It has been demonstrated that microbially deposited manganese oxide on stainless steel (UNS S31603) (Dickinson and Lewandowski, 1996) in freshwater caused an increase in E_{corr} and increased cathodic current density at potentials above −200 mV (vs. SCE). For carbon steel corrosion under anodic control, the oxides can elevate corrosion current, but will cause little positive shift in E_{corr}. The increase in corrosion current may be significant, particularly for carbon steel covered with biomineralized oxides and corrosion product tubercles that provide large mineral surface areas. Given sufficient conductivity in the tubercle, much of this material may serve as an oxide cathode to support corrosion at the oxygen-depleted anode within the tubercle. Continued biomineralization within a large tubercle may sustain a significant amount of the cathodic current. Both factors can increase the risk of corrosion. Ennobled E_{corr} can exceed pitting potentials for low molybdenum stainless-steel alloys, enhancing risk of pit nucleation, while elevated cathodic current density impedes repassivation. Biomineralized manganic oxides are efficient cathodes and increase cathodic current density on stainless steel by several decades at potentials between approximately -200 and $+400$ mV$_{SCE}$. The extent to which the elevated current density can be maintained is controlled by the electrical capacity of the mineral reflecting both total accumulation and conductivity of the mineral–biopolymer assemblage (only material in electrical contact with the metal will be cathodically active). The biomineralization rate and the corrosion current control oxide accumulation in that high corrosion currents will discharge the oxide as rapidly as it is formed. This variation in accumulation causes the oxides to exert different modes of influence on the corrosion behavior of active metals compared with passive metals.

Dickinson and Pick (2002) proposed the following three thresholds for potential manganese deposition: (1) waters with less than 10 ppb Mn(II)—low potential, (2) between 10 and 50 ppm Mn(II)—deposition may occur if manganese-oxidizing microorganisms or oxidizing biocides are present, and (3) greater than 50 ppb Mn(II)—potential high.

Iron

The iron-oxidizing genera that are usually cited as causing MIC are *Gallionella*, *Sphaerotilus*, *Crenothrix*, *Siderocapsa*, *Clonothrix*, and *Leptothrix*. These organisms oxidize ferrous ions to ferric ions or manganous ions to manganic ions to obtain energy. Iron-depositing bacteria produce orange-red tubercles of iron oxides and hydroxides by oxidizing ferrous ions from the bulk medium or the substratum (Figure 2-18). With use of environmental scanning electron microscopy (ESEM), it is possible to locate iron-depositing bacteria with tubercles, twisted filaments with iron-rich deposits along their length (Figure 2-19*a*, *b*). Iron-oxidizing bacteria are microaerophilic and may require synergistic associations with other bacteria to

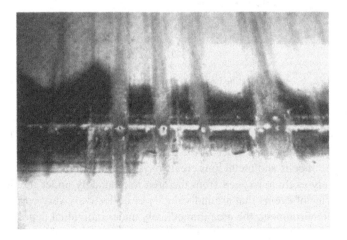

FIGURE 2-18   Tubercles produced by iron-depositing bacteria associated with 316L stainless-steel (UNS S31603) weld. (Kobrin, 1976. © NACE International, 2003.)

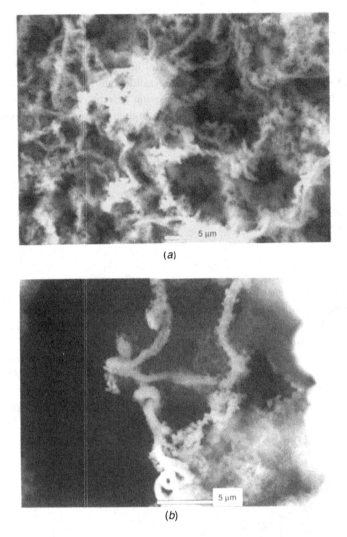

(a)

(b)

FIGURE 2-19a, b   Iron-depositing bacteria within turbercles. (Little et al., 1998.)

maintain microaerophilic conditions in their immediate environment. Bacilli and cocci were imaged on the surface of the tubercle.

Deposits of cells and metal ions create oxygen concentration cells (Figure 2-2) that effectively exclude oxygen from the area immediately under the deposit and initiate a series of events that are individually or collectively very corrosive. In an oxygenated environment, the area immediately under individual deposits becomes deprived of oxygen. This area becomes a relatively small anode compared to the large surrounding oxygenated cathode. Corrosion at the anode produces metal ions that hydrolyze and decrease the pH. In addition, Cl^- ions from the electrolyte will migrate to the anode to neutralize any buildup of charge, forming heavy-metal chlorides that are extremely corrosive. Under these circumstances, pitting involves the conventional features of differential aeration, a large cathode–anode surface area, and the development of acidity and metallic chlorides. Pit propagation is dependent not on activities of the organisms but on metallurgy (George, 1996). This also means that attempts to kill the organisms within mineral deposits using biocides will not result in a cessation of pit propagation (Miller and Tiller, 1970). Stainless steels containing 6 percent or more molybdenum are not vulnerable to this type of attack.

Metal Reduction

Dissimilatory iron and/or manganese reduction occurs in several microorganisms, including anaerobic and facultative aerobic bacteria. Inhibitor and competition experiments suggest that Mn(IV) and Fe(III) are efficient electron acceptors similar to nitrate in redox ability and are capable of outcompeting electron acceptors of lower potential, such as sulfate or carbon dioxide (Myers and Nealson, 1988). Many of the recently described metal-reducing bacteria are capable of using a variety of electron acceptors, including nitrate and oxygen (Myers and Nealson, 1988). Myers and Nealson (1989) suggested that iron and manganese-reducing microorganisms must be in direct contact with oxides to reduce them. This conclusion is based on the observation that Fe(III) and Mn(IV) are not reduced if microorganisms capable of the reduction are separated from the oxides by a semipermeable membrane that allows exchange of soluble molecules but prevents contact between the organism and the oxide. These experiments demonstrated that metal reduction by *Shewanella putrefaciens* (renamed *oneidensis*) required cell/surface contact and the rate of reduction was directly related to surface area. Environmental scanning electron microscopy and confocal laser scanning microscopy (CLSM) images of bacteria during both the iron and manganese reduction processes showed close contact of the cells with the oxides during the early stages of reduction. In later stages, manganese oxides were coated with a layer of extracellular material that obscured the cells (Little et al., 1997).

Little et al. (1997) used synthetic iron oxides, oxyhydroxides [goethite, α-FeO·OH; hematite, Fe_2O_3; and ferrihydrite, $Fe(OH)_3$], as model compounds to simulate the mineralogy of passivating films on carbon steel. There is general agreement that oxide films formed on iron in air at temperatures below 200 °C are composed of magnetite and hematite. Szklarska-Smialowska (1986) described the formation of hematite over a magnetite film. Ferric oxyhydroxides, including goethite and lepidocrocite (γ-FeO·OH), have also been identified in protective layers on carbon steel. Under anaerobic conditions goethite, hematite, and ferrihydrite were reduced by *S. oneidensis* (Table 2-2). Rates of reduction, measured by atomic absorption spectroscopy of Fe(II) in solution as a function of time, for the three minerals indicate that after a 24-h exposure to *S. oneidensis*, initial reduction rates for goethite and ferrihydrite were approximately the same

TABLE 2-2 Relative Rates of Iron Reduction Based on Fe(II) in Solution Exposed to *Shewanella oneidensis*

	Twenty four hours (mg L^{-1} d^{-1})	% Max[a]	Twenty two days (mg L^{-1} d^{-1})	% Max
Goethite	3.50	100	18.20	100
Ferrihydrite	2.60	74	14.10	77
Hematite	0.58	22	0.39	2

Source: Little et al. (1997).
[a]Max percentage indicates rates normalized to goethite reduction.

and were five times faster than the reduction rate for hematite. After 22 days, the integrated reduction rates for goethite and ferrihydrite were much faster than those measured at 24 h. The hematite reduction rate actually slowed over the exposure period so that after 22 days the overall integrated rate was 50 times slower than reduction rates for goethite and ferrihydrite. Obuekwe et al. (1981) observed that *Pseudomonas* sp. (later renamed *Shewanella* sp.), an iron-reducing bacterium, accelerated the corrosion of 1018 steel (UNS G10180). In the presence of bacteria, a surface film formed that was not passivating.

Methane Production

Methanogenic *Archaea* (methanogens) are obligate anaerobes. Several strains of methanogens (e.g., *Methanosarcina barkeri*, *Methanobacterium bryantii*, and *Methanospirillum hungatei*) can grow in media containing iron or other metals as the only source of electrons (Daniels et al., 1987; Belay and Daniels, 1990). The result is methane production and an increase in Fe^{2+} formed from metallic iron. Although methane production with metallic iron has been demonstrated, acceleration of corrosion due to this process has not been shown.

Hydrogen Production

Walch and Mitchell (1983) proposed the following roles for microorganisms in hydrogen embrittlement of metals: (1) production of molecular hydrogen during fermentation, which may be dissociated into atomic hydrogen and absorbed into metals; (2) production of hydrogen ions via organic or mineral acids, which may be reduced to form hydrogen atoms at cathodic sites; (3) production of H_2S, which stimulates the absorption of atomic hydrogen into metals by preventing its recombination into hydrogen molecules; and (4) destabilization of metal-oxide films.

Dealloying

Biofilm formation of copper alloys often results in selective dealloying. Zinc tends to be selectively removed from copper–zinc alloys, producing a spongy copper material (Walker, 1977) through the incorporation of zinc in corrosion products to produce rosasite ($(Cu,Zn)_2CO_3(OH)_2$) (Gettens, 1963, 1969). Dealloying of the copper from numerous tin bronzes has been reviewed by Geilmann (1956). Partitioning of alloying elements between remaining metal ions in the corrosion product and the electrolyte has received little attention (Zolotarev et al., 1987).

INACTIVATION OF CORROSION INHIBITOR

Biofilms reduce the effectiveness of corrosion inhibitors by creating a diffusion barrier between the metal surface and the inhibitor in the bulk solution (Kirkpatrick et al., 1980). Futhermore, many of the compounds used as corrosion inhibitors can provide nutrients. Aliphatic amines and nitrites used as corrosion inhibitors can be degraded by microorganisms, decreasing the effectiveness of the compounds and increasing the microbial populations (Pope, 1987; Pope et al., 1984a,b). Cooke et al. (1995) reported that potassium chromate (K_2CrO_4) was ineffective as a corrosion inhibitor in an electricity generating station because of chromate-reducing bacteria. The bacteria were causing blockage of pipes by precipitation of chromium(III) oxide.

ALTERATION OF ANION RATIOS

Microorganisms can also alter the electrolyte and make it more corrosive. Molar ratios of aggressive ions to inhibiting ions (e.g., $2 Cl^-$ to $\Sigma NO_3^- + SO_4^{2-} + PO_4^{3-}$) are used to predict whether an electrolyte can sustain a localized corrosion reaction. The relationships between the concentration of inhibitive and aggressive anions correspond to competitive uptake of the anions by adsorption or ion exchange at a fixed number of sites at the oxide surface. Increasing chloride concentration shifts the critical pitting potential to more active (negative) values. The potential is shifted to more noble (positive) values by the presence of other anions, particularly oxyanions (ClO_4^-, SO_4^{2-}, NO_3^-, PO_4^{3-}, NO_2^-, and OH^-). Kehler et al. (2001) concluded that electrolyte composition and molar ratios of anions had a stronger effect on crevice stabilization potentials than bulk pH at temperatures between 60 and 95 °C for both alloy 625 (UNS N06625) and alloy 22 (UNS N06022). However, microorganisms can consume oxyanions. The concentrations and types of anions required for corrosion inhibition are extremely specific for both metals and environments. For example, sulfates can inhibit chloride-induced pitting of stainless steels, but are aggressive toward carbon steel. Sulfates break down passivity in nitrite solutions more than chlorides.

To be fully effective inhibitors, anions must be present in certain minimum concentrations. In many applications, excursions in solution chemistry or temporary loss of inhibitor may give rise to localized corrosion in an otherwise inhibited system (Turnbull et al., 2002). At concentrations below the critical value, inhibitive anions may act aggressively and stimulate breakdown of oxide films. Not only is initial concentration important, but also concentration during service. Salvarezza et al. (1983) used pitting potential to assess MIC of an aluminum alloy (UNS A92024) in relation to the electrolyte composition. During growth of the fungus *Cladosporium* (renamed *Hormoconis*) *resinae*, nitrate and phosphate were incorporated into the biomass, increasing the chloride/inhibitor ratio. In their experiments, fungal uptake of inhibitors was the principal cause of the pitting potential decrease during microbial growth. Predictions about pitting or crevice corrosion resistance in chloride-containing media cannot be based on anion ratios without consideration of potential microbial alterations (Little, 2003).

SUMMARY

There are many mechanisms for microbiologically influenced corrosion and many causative microorganisms for each mechanism. The most aggressive microbiologically influenced corrosion takes place in the presence of microbial consortia in which many physiological types of bacteria interact in complex ways within biofilms. The list of specific microorganisms responsible for microbiologically influenced corrosion is constantly growing as detection techniques improve and new mechanisms are identified. However, the following groups of bacteria are consistently identified as causative organisms: slime-producing, sulfur-oxidizing, sulfate-reducing, metal-oxidizing, and metal-reducing bacteria, in addition to acid-producing bacteria and fungi.

REFERENCES

Arnold RG, DiChristina T, Hoffmann M (1988). Reductive dissolution of Fe(III) oxides by *Pseudomonas* sp. 200. *Biotechnol. Bioeng.*, **32**: 1081–1096.

Baas-Becking GM, Moore D (1961). Biogenic sulfides. *Econ. Geol.*, **56**: 259–272.

Banister FA (1952). *An Unusual Synthesis of Acanthite Crystals*. Mineralogical Society of London Meeting, Reston, VA. Ford-Fleischer files of the U.S. Geological Survey.

Bauer R (1988). *Sulfide Corrosion of Silver Contacts during Satellite Storage*. U.S. Air Force Report; SD-TR-88-53/AD-1196 217.

Belay N, Daniels L (1990). Elemental metals as electron source for biological methane formation from CO_2, *Antonie van Leeuwenhoek*, **57**: 1–7.

Berner RA (1969). The synthesis of framboidal pyrite. *Econ. Geol.*, **64**: 383–384.

Blunn G (1986). Biological fouling of copper and copper alloys. *Biodeterior.*, **6**: 567–575.

Booth GH, Tiller AK (1962). Polarization studies of mild steel in cultures of sulphate reducing bacteria. *Trans. Faraday Soc.*, **58**: 2510.

Burns JM, Staffeld EE, Calderon OH (1969). Corrosion caused by organic acids reacting with three metals with special emphasis on Krebs citric acid cycle acids. *Dev. Ind. Microbiol.*, **8**: 327–331.

Characklis WG, Marshall KC (1990). Biofilms: a basis for an interdisciplinary approach. In: WG Characklis, KC Marshall (eds) *Biofilms*. New York: Wiley, p. 4.

Chiou W A, Kohyama N, Little B J, Wagner P A, Meshii M (1996). TEM study of a biofilm on copper corrosion. In: Bailey G W, Corbett J M, Dimlich R V W, Michael J R and Zaluzec N J (eds) *Proceedings of Microscopy and Microanalysis*. San Francisco: San Francisco Press, Inc., pp. 220–221.

Cooke VM, Hughes MN, Poole RK (1995). Reduction of chromate by bacteria isolated from the cooling water of an electricity generating station. *J. Ind. Microbiol.*, **14**: 323–328.

Costerton JW, Geesey GG (1986). The microbial ecology of surface colonization and of consequent corrosion. In: Dexter SC (ed) *Biologically Induced Corrosion*. Houston, TX: NACE International, pp. 223–232.

Craig JR, Scott SD (1974). Sulfide phase equilibria. In: PH Ribbe (ed)*Sulfide Mineralogy*. Washington, DC: Mineralogical Society of America, pp. —CS1–CS110.

Crolet JL, Magot M (1995). Observations of non-SRB sulfidogenic bacteria from oilfield production facilities. *CORROSION/1995*, Paper No. 95188. Houston, TX: NACE International.

Cuthbert M (1962). Formation of bornite at atmospheric temperature and pressure. *Econ. Geol.*, **57**: 38–41.

Daniels L, Belay N, Rajagopal BS, Weimer PJ (1987). Bacterial methanogenesis and growth from CO_2 with elemental iron as the sole source of electrons. *Science*, **237**: 509–511.

Daubree GA (1862). Contemporary formation of copper pyrite by the action of hot springs at Bagnes-de-Bigorre. *Bull. Soc. Geol. France*, **19**: 529–532.

de Gouvernain M (1875). Sulfiding of copper and iron by a prolonged stay in the thermal spring at Bourbon l'Archambault. *Compte Rendu*, **80**: 1297–1300.

Dexter SC (2003). Microbiologically influenced corrosion. In: Cramer SD, Covino Jr. BS (eds) *Corrosion: Fundamentals, Testing and Protection, ASM Handbook*. ASM International, Metals Park, OH Vol. 13A, p. 400.

Dickinson WH, Lewandowski Z (1996). Manganese biofouling and the corrosion behavior of stainless steel. *Biofouling*, **10**: 79–93

Dickinson WH, Caccavo F, Lewandowski Z (1996). The ennoblement of stainless steel by manganic oxide biofouling. *Corros. Sci.*, **38**: 1407–1422.

Dickinson WH, Pick PW (2002). Manganese-dependent corrosion in the electric utility industry. *CORROSION/2002*, Paper No. 02444, Houston, TX: NACE International.

Dinh, HT, Kuever J, Mumann M, Hassel AW, Stratmann M, Widdel F (2004). Iron corrosion by novel anaerobic microorganisms. *Nature*, **427**: 829–832.

Duncan SJ, Ganiaris H (1987). Some sulphide corrosion products on copper alloys and lead alloys from London waterfront sites. In: Black EJ (ed) *Recent Advances in the Conservation and Analysis of Artifacts*. London: Summer Schools Press, University of London, pp. 109–118.

Evans HT (1979). The crystal structures of low chalcocite and djurleite. *Z . Krist.*, **150**: 299–320.

Ford TE, Black JP, Mitchell R (1990). Relationship between bacterial exopolymers and corroding metal surfaces. *Proceedings of Corrosion '90*, Las Vegas, Paper No. 110, National Association of Corrosion Engineers.

Gaylarde CC, Videla HA (1987). Localized corrosion induced by a marine *vibrio. Int. Biodeterior.*, **23**(2): 91–104.

Garrels RM, Christ JC (1965). *Solutions, Minerals, and Equilibria*. San Francisco: Freeman Cooper.

Geesey GG, Mittleman MW, Iwaoka T, Griffiths PR (1986). Role of bacterial exopolymers in the deterioration of metallic copper surfaces. *Mater. Perform.*, **25**(2): 37–40.

Geesey GG, Bremer PJ (1999). Application of Fourier transform infrared spectrometry to studies of copper corrosion under bacterial colonies. *Marine Tech. Soc. J.*, **24**(3): 36–42.

Geilmann W (1956). Leaching of bronzes in sand deposits. *Angew. Chem.*, **68**: 201–211.

George RP (1996). Studies on the corrosion and tuberculation of carbon steel in fresh water. *National Symposium of Research Scholars on Metals and Materials Research.* Chennai, India: Institute of Metals, pp. 118–127.

Gerchakov SM, Udey LL (1984). *Marine Biodeterioration: An Interdisplinary Study.* Annapolis: Naval Institute Press; p. 83.

Gettens RJ (1963). *The Corrosion Products of Metal Antiquities.* Annual Report to the Trustees of the Smithsonian Institution. Washington, DC: Smithsonian Institution, pp. 547–568.

Gettens, RJ (1969). *The Freer Chinese Bronzes, Vol. 2, Technical Studies.* Washington, DC: Smithsonian Institution, p. 257.

Ghiorse WC (1984). Biology of iron- and manganese-depositing bacteria. *Annu. Rev. Microbiol.*, **38**: 515–550.

Goldschmidt J (1953). A simplexity principle. *J. Geol.*, **61**: 539–551.

Gounot AM (1994). Microbial oxidation and reduction of manganese: consequences in groundwater and applications. *FEMS Microbiol. Rev.*, **14**: 339–350.

Hack H, Shih H, Pickering HW (1986). Role of the corrosion product film in the corrosion protection of Cu-Ni alloys in seawater. In: McCafferty E, Broadd RJ (eds) *Surfaces, Inhibition, and Passivation.* Pennington, NJ: The Electrochemical Society, pp. 355–367.

Hamilton WA (1985). Sulphate-reducing bacteria and anaerobic corrosion. *Annu. Rev. Microbiol.*, **39**: 195–217.

Hardy JA, Brown JL (1984). Corrosion of mild steel by biogenic sulfide films exposed to air. *Corrosion*, **40**: 650–654.

Heimann RB (1989). Assessing the technology of ancient pottery: the use of ceramic phase diagrams. *Archeomaterials*, **31**: 123–148

Johnsen R, Bardal E (1985). Cathodic properties of different stainless steels in natural seawater. *Corrosion*, **41**(5): 296–302.

Jung WK, Schweisfurth R (1976a). Manganoxydierende bakterien. II. *Z. Allg. Mikrobiologie*, **16**(2): 133–147.

Jung WK, Schweisfurth R (1976b). Manganoxydierende bakterien. III. *Z. Allg. Mikrobiologie*, **16**(8): 587–597.

Kehler BA, Ilevbare GO, Scully JR (2001). Crevice corrosion stabilization and repassive behavior of alloy 625 and alloy 22. *Corrosion*, **57**(12): 1042–1065.

King RA, Miller JDA, Wakerly DS (1973). Corrosion of mild steel in cultures of sulphate-reducing bacteria; effect of changing the soluble iron concentration during growth. *Br. Corros. J.*, **8**: 89–93.

Kirkpatrick JP, Mcintire LV, Characklis WG (1980). Mass and heat transfer in circular tube with biofouling. *Water Res.*, **14**: 117.

Kobrin G (1986). Reflections in microbiologically induced corrosion of stainless steels. In: Dexter SC (ed) *Biologically Induced Corrosion.* Houston, TX: NACE International, p. 37.

Kostov I, Minceva-Stefanova J (1981). *Sulphide Minerals.* Sofia, Bulgaria: Publishing House of the Bulgarian Academy of Sciences.

Laque FL (1975). *Marine Corrosion.* New York: Wiley.

Lee W, Lewandowski Z, Morrison M, Characklis WG, Avci R, Nielsen PH (1993). Corrosion of mild steel underneath aerobic biofilms containing sulfate-reducing bacteria part II: at high dissolved oxygen concentration. *Biofouling*, **7**: 217–239.

Little B, Wagner P, Gerchakov SM, Walch M, Mitchell R (1986). Involvement of a thermophilic bacterium in corrosion processes. *Corrosion*, **42**(9): 533.

Little B, Wagner P, Duquette D (1988). Microbiologically induced increase in corrosion density of stainless steel under cathodic protection. *Corrosion*, **44**(5): 270.

Little, BJ (2003). A perspective on the use of anion ratios to predict corrosion in Yucca Mountain. *Corrosion*, **59**(8): 701–704.

Little BJ, Wagner PA, Lewandowski Z (1998). The role of biomineralization in microbiology influenced corrosion. *CORROSION/1998*, Paper No. 294, Houston, TX: NACE International.

Little BJ, Wagner PA, Ray RI (1991). Microbiologically influenced corrosion of copper alloys in saline waters containing sulfate-reducing bacteria. *CORROSION/1991*, Paper No. 101, Houston, TX: NACE International.

Little B, Wagner P, Hart K, Ray R, Lavoie D, Nealson K, Aguilar C (1997). The role of metal-reducing bacteria in microbiologically influenced corrosion. *CORROSION/1997*, Paper No. 215, Houston, TX: NACE International.

Lucey VF (1967). Mechanism of pitting corrosion of copper in supply waters. *Br. Corros. J.*, **2**: 175–185.

Macdonald DD, Syrett BC, Wing SS (1979). Corrosion of Cu–Ni alloys 706 and 715 in flowing sea water – 2. Effect of dissolved sulfide. *Corrosion*, **35**: 367–378.

Mansfeld F, Little B (1991). A technical review of electrochemical techniques applied to microbiologically influenced corrosion. *Corros. Sci.* **32**(3): 247–272.

McNeil MB, Mohr DW (1992). Interpretation of bronze disease and related copper corrosion mechanisms in terms of log activity diagrams. In: Scott DA (ed) *Materials Problems in Art and Archaeology II*. Pittsburgh, PA: Materials Research Society, pp. 1055–1063.

McNeil MB, Odom AL (1994). Thermodynamic predictions of microbiologically influenced corrosion by sulfate-reducing bacteria. In: Kearns JR, Little BJ (eds) *Microbiologically Influenced Corrosion Testing STP 1232*. Fredericksburg, VA: American Society for Testing Materials, pp. 173–179.

McNeil M, Little BJ (1992). Corrosion mechanisms for copper and silver objects in near surface environments. *J. Am. Inst. Conserv.*, **31**(3): 355–366.

McNeil M, Little BJ (1999). The use of mineralogical data in interpretation of long-term microbiological corrosion processes: sulfiding reactions. *J. Am. Inst. Conserv.*, **38**: 186-199.

McNeil MB, Mohr DW (1993). Formation of copper-iron sulfide minerals during corrosion of artifacts and implications for pseudogilding. *Geoarchaeology*, 8, 1:23–33.

McNeil MB, Jones J, Little BJ (1991). Mineralogical fingerprints for corrosion processes induced by sulfate-reducing bacteria. *CORROSION/91;*, Paper No. 580, Houston, TX: NACE International.

McNeil MB, Little BJ (1990). Mackinawite formation during microbial corrosion. *Corrosion*, **46**(7): 599–600.

Miller JDA, Tiller AK (1970). Microbial corrosion of buried and immersed metal. In: Miller JDA (ed) *Microbial Aspects of Metallurgy*. New York: American Elsevier, pp. 61–106.

Mond L, Cuboni G (1893). On the nature of antique bronze patina. *Atti. Reale Accad. dei Lincei Serie*, **52**: 498–499.

Mor ED, Beccaria AM (1975). Behaviour of copper in artificial seawater containing sulphides. *Br. Corros. J.*, **10**: 33–38.

Myers C, Nealson KH (1988). Bacterial manganese reduction and growth with manganese oxide as the sole electron acceptor. *Science*, **240**: 1319–1321.

Nealson K, Tebo B, Rosson R. Occurrence and mechanisms of microbial oxidation of manganese (1988). In: A Laskin (ed) *Advances in Applied Microbiology*. New York: Academic Press, Vol. 33. pp. 299–318.

Nilsson I, Ohlson S, Haggstrom L, Molin N, Mosbach K (1980). Denitrification of water using immobilized *Pseudomonas denitrificans* cells. *Eur. J. Appl. Microbiol. Biotechnol.*, **10**: 261–274.

Nivens DE, Nichols, PD, Henson, JM, Geesy GG, White DC (1986). Reversible acceleration of the corrosion of AISI 304 stainless stell exposed to seawater induced by growth and secretions of the marine bacterium *Vibrio natriegens*. *Corrosion*, **42**(4): 204.

North NA, MacLeod ID. Corrosion of metals (1986). In: Pearson C (ed) *Conservation of Archaeological Objects*. London: Butterworths; pp. 69–98.

North RF, Pryor MJ (1970). The influence of corrosion product structure on the corrosion rate of Cu–Ni alloys. *Corros. Sci.*, **10**: 297–311

Obuekwe CO, Westlake DWS, Plambeck JA, Cook FD (1981). Corrosion of mild steel in cultures of ferric iron reducing bacterium isolated from crude oil I. polarization characteristics. *Corrosion*, **37**(8): 461–467.

Pfennig N, Widdel F, Truper HG (1981). The dissimulatory sulfate-reducing bacteria. In: Starr MP, Stolp M, Truper HG, Balows A, Schlegel HG (eds) *The Prokaryotes: A Handbook on Habitats*. New York: Springer-Verlag, pp. 926–940.

Pollard AM, Thomas RG, Williams PA (1989). Synthesis and stabilities of basic copper (II) chlorides atacamite, paratacamite, and botallackite. *Mineralog. Mag.*, **53**: 557–563.

Pope DH, Duquette DJ, Johannes AH, Wayner PC (1984a). Microbiologically influenced corrosion of industrial alloys. *Mater. Perform.*, **23**(4): 14–18.

Pope DH, Duquette DJ, Wayner PC, Johannes AH (1984b). *Microbiologically Influenced Corrosion: A State of the Art Review*. Columbus, OH: Materials Technology Institute of the Chemical Process Industries.

Pope DH (1987). *Microbial Corrosion in Fossil-Fired Power Plants*. Electric Power Research Institute, CS-5495, Palo Alto, CA., pp. 6–3.

Postgate JR (1979). *The Sulphate-Reducing Bacteria*. Cambridge, UK: Cambridge University Press.

Pourbaix M (1966). Atlas of electrochemical equilibria in aqueous solutions. Houston, TX: NACE International.

Putnis A (1977). Electron diffraction study of phase transformations in copper sulfides. *Am. Mineralog.*, **62**: 107–114.

Ribbe PH (ed) (1976). *Sulfide Mineralogy*. Washington, DC: Mineralogical Society of America.

Roseboom EH (1966). An investigation of the system Cu-S and some natural copper sulfides between 25 degrees and 700 degrees. *Econ. Geol.*, **61**: 641–672.

Salvarezza RC, de Mele FL, Videla HA (1983). Mechanisms of the microbial corrosion of aluminum alloys. *Corrosion*, **39**: 26–32.

Schweisfurth R, Heitz E (1989). *Mikrobiologische Materialzerstoerung und Materialschutz*. Frankfurt am Main, West Germany: DECHEMA.

Scott DA (1990). Bronze disease; a review of some chemical problems and the role of relative humidity. *J. Am. Inst. Conserv.*, **29**: 192–306.

Shrier LL, Jarman RA, Burstein GT (eds) (1994). *Corrosion* (3rd edn). Oxford: Butterworth-Heinemann Ltd., Vols. 1, 2.

Staffeldt EE, Kohler DA (1973). *Assessment of Corrosion Products Removed from "La Fortuna."* Punta del Mar, Venezia: Petrolia e Ambiente, pp. 163–170.

Stranger-Johannessen M (1986). Fungal corrosion of the steel interior of a ship's holds. In: *Biodeterioration VI.* Slough, UK: CAB Int., pp. 218–223.

Syrett BC (1977). Accelerated corrosion of copper in flowing pure water contaminated with oxygen and sulfide. *Corrosion*, **33**: 257–262.

Syrett BC (1980). The mechanism of accelerated corrosion of copper–nickel alloys in sulfide polluted seawater. *CORROSION/1980*, Paper No. 33, Houston, TX: NACE International.

Syrett BC (1981). The mechanism of accelerated corrosion of copper-nickel alloys in sulphide-polluted seawater. *Corros. Sci.*, **21**: 187–209.

Szklarska-Smialowska Z (1986). Pitting corrosion of metals. Houston, TX: NACE International. In: Terry LA, Endyvean RGJ (eds) *Algal Biofouling.* New York: Elsevier, p. 211.

Terry LA, Edyvean RGJ (1986). Recent investigations into the effects of algae on corrosion. In: Evans LV, Hoagland KD (eds) *Algal Biofouling – Studies in Environmental Science.* New York: Elsevier, Vol. 8.

Turnbull A, Coleman D, Griffiths A, Francis PE, Orkney L (2002). Effectiveness of corrosion inhibitors in retarding propagation of localized corrosion. *CORROSION/2002*, Paper No. 274, Denver, CO.

Uhlig HU, Revie RW (1985). *Corrosion and Control, An Introduction to Corrosion Science and Engineering.* New York: Wiley.

von Wolzogen Kuhr CAH, van der Vlugt LS (1934). *Water, The Hague*, **18**: 147.

Wagman DD, Evans WH, Parker VB, Schumm RH, Halow I, Bailey SM, Churney KL, Nuttall RL (1982). The Nbs tables of chemical thermodynamic properties: selected values for inorganic and C_1 and C_2 organic substances in SI units. *J. Phys. Chem.*, Reference Data 11, Supplement 2.

Walch M, Mitchell R (1983). The role of microorganisms in hydrogen embrittlement of metals. *CORROSION/1983*, Paper No. 249, Anaheim, CA: NACE International.

Walker GD (1977). An SEM and microanalytical study of in-service dezincification of brass. *Corrosion*, **33**: 252–256.

Wikjord AG, Rummery TE, Doern FE, Owen DG (1980). Corrosion and deposition during the exposure of carbon steel to hydrogen sulfide water solutions. *Corros. Sci.*, **20**: 651–671.

Zolotarev EI, Pchel'mikov AP, Skuratnick YaB, Dembrovskii MA, Khokhlov NI, Losev VV (1987). Kinetics of dissolution of copper–nickel alloys, anodic dissolution of Cu–30% Ni under steady state conditions. *Zashchita Metallov*, **23**: 922–929.

Chapter 3

Diagnosing Microbiologically Influenced Corrosion

INTRODUCTION

Diagnosing microbiologically influenced corrosion (MIC) after it has occurred requires a combination of microbiological, metallurgical, and chemical analyses. Microbiologically influenced corrosion investigations have typically attempted to (1) identify causative microorganisms in the bulk medium or associated with the corrosion products, (2) identify a pit morphology consistent with an MIC mechanism, and (3) identify a corrosion product chemistry that is consistent with the causative organisms. The following sections provide a discussion of available techniques, their advantages and disadvantages, and, most importantly, their limitations.

IDENTIFICATION OF CAUSATIVE ORGANISMS

For many years, the first step in identifying corrosion as an MIC was to determine the presence of specific groups of bacteria in the bulk medium (planktonic cells) or those associated with corrosion products (sessile cells). There are four approaches: (1) culture the organisms on solid or in liquid media, (2) extract and quantify a particular cell constituent, (3) demonstrate/measure some cellular activity, or (4) demonstrate a spatial relationship between microbial cells and corrosion products using microscopy.

Culture Techniques

The method most often employed for detecting and enumerating groups of bacteria is the serial dilution to extinction method, using selective culture media. To culture microorganisms, a small amount of liquid or a suspension of a solid (the inoculum)

Microbiologically Influenced Corrosion By Brenda J. Little and Jason S. Lee
Published 2007 by John Wiley & Sons, Inc.

is added to a solution or solid containing the nutrients (culture medium). The following three factors are considered when growing microorganisms: type of culture medium, incubation temperature, and length of incubation. The present trend in culture techniques is to attempt to culture several physiologically divergent groups, including aerobic bacteria; facultative anaerobic bacteria; sulfate-reducing bacteria (SRB); and acid-producing bacteria (APB). Growth is detected as turbidity or a chemical reaction within the culture medium.

Traditional SRB media contain sodium lactate as the carbon source (API, 1965; Postgate, 1979). When SRB are present in the sample, sulfate is reduced to sulfide, which reacts with iron (either in solution or solid) to produce black ferrous sulfide. Culture media are typically observed over several days (30 days may be required for the growth of SRB). There have been several attempts to improve the culture media to grow higher numbers of bacteria or shorten the time required for some indication of growth. A complex SRB medium was developed containing multiple carbon sources that could be degraded into both acetate and lactate. In comparison tests, the complex medium produced higher counts of SRB from waters and surface deposits among five commercially available media (Scott and Davies, 1992). Jhobalia et al. (2005) developed an agar-based culture medium for accelerating the growth of SRB. The authors noted that over the range 1.93–6.50 g L^{-1}, SRB grew best at the lowest sulfate concentration. Cowan (2005) developed a rapid culture technique for SRB based on rehydration of dried nutrients with water from the system under investigation. The author claimed that using system water reduced the acclimation period for microorganisms by ensuring that the culture medium had the same salinity as the system water used to prepare the inoculum. The author reported quantification of SRB within 1 to 7 days.

A distinct advantage of culturing techniques used to detect specific microorganisms is that low numbers of cells can be grown to easily detectable higher numbers in the proper culture medium. However, there are numerous limitations for the detection and enumeration of cells by culturing techniques. Several investigators have followed the changes in microflora as a function of water storage. Zobell and Anderson (1936) and Lloyd (1937) demonstrated that when water is stored in glass bottles, bacterial numbers fall within the first few hours, followed by an increase in the total bacterial population and a reduction in the number of species. If results from culturing techniques are to be related to natural populations, the culture media should be inoculated within hours of collection and the sample chilled during the interim. Under all circumstances, culture techniques underestimate the organisms in a natural population (Giovanni et al., 1990; Ward et al., 1998). Kaeberlein et al. (2002) suggest that 99 percent of microorganisms from the environment resist cultivation in the laboratory.

A major problem in assessing microorganisms in natural environments is that viable microorganisms can enter into a nonculturable state (Rosak and Colwell, 1987). Another problem is that the culture media cannot approximate the complexity of a natural environment. Growth media tend to be strain-specific. For example, lactate-based media sustain the growth of lactate oxidizers, but not acetate-oxidizing bacteria. Incubating at one temperature is further selective. The type of medium used to culture microorganisms determines to a large extent the numbers and types of microorganisms that grow. Zhu et al. (2005) demonstrated dramatic changes in the microbial population from a gas pipeline after samples were introduced into liquid culture media. For example, SRB dominated the microflora in most pipeline samples when culture techniques were used. However, using culture-independent genetic techniques, they found that methanogens were more abundant in most pipeline samples than denitrifying bacteria and that SRB were the least abundant bacteria. Similarly, Romero et al. (2005) used genetic monitoring to identify bacterial populations in a seawater injection system. They found that some bacteria, present in small amounts in the original waters, were enriched in the culture process.

Biochemical Assays

Biochemical assays have been developed for the detection of specific microorganisms associated with MIC. Unlike culturing techniques, biochemical assays for detecting and quantifying bacteria do not require growth of the bacteria. Instead, these assays measure constitutive properties, including adenosine triphosphate (ATP) (ASTM, 1977), phospholipid fatty acids (PLFA) (Franklin and White, 1991), cell-bound antibodies (Pope, 1986), and DNA (Hogan, 1990). Adenosine-5'-phosphosulfate (APS) reductase (Tatnall et al., 1988) and hydrogenase (Boivin et al., 1990) have been used to estimate SRB populations.

Since ATP is a compound found in all living matter, ATP assays estimate the total number of viable organisms in a sample. ATP assays are based on the luciferin–luciferase reaction, where ATP provides energy for the oxidation of luciferin by the enzyme luciferase. The procedure requires that a water sample be filtered to remove solids and salts, which may interfere. The filtered sample is added to a reagent to release cell ATP. The reaction is sensitive to sulfide, some metals, and some types of biocides. Emitted light is measured with a photometer, and the amount of light released during the reaction is directly related to the amount of ATP in the sample.

Biofilm community structure can be analyzed using cluster analysis of phospholipid fatty acid profiles (Franklin and White, 1991). Phospholipids are found in the membranes of all cells. Under the conditions in natural communities, bacteria contain a relatively constant proportion of their biomass as phospholipids. Phospholipids are not found in storage lipids and have a relatively rapid turnover so that their assay gives a measure of the viable cellular biomass. The phosphate of phospholipids or the glycerol-phosphate and acid-labile glycerol from phosphatidyl glycerol-like lipids can be assayed to increase the specificity and sensitivity of the phospholipids assay. The ester-linked fatty acids in phospholipids are both the most sensitive and the most useful chemical measures of microbial biomass and community structure. Phospholipid fatty acid profiles for natural biofilms have been shown to be more complex than profiles for laboratory biofilms (Franklin and White, 1991). None of the laboratory profiles clustered closely with profiles from natural biofilms. In addition, phospholipid fatty acid profiles for attached bacteria clustered separately from profiles of the same bacteria in bulk phase, suggesting that either the community or physiology of the attached bacteria differs from that of bulk phase bacteria. Despite the fact that PLFA analysis cannot provide an exact description of each species in a given environment, the analysis provides a quantitative description of the microbiota in a particular environment. Analysis of other components of the phospholipids fraction gives insights into community structure. For example, SRB contain lipids that can be used to identify at least some species. Dowling et al. (1988) identified unusual fatty acids as biomarkers for two SRB: iso 17:1w7c and branched monoenoics for a hydrogen-oxidizing *Desulfovibrio* sp. and 10 methyl 16:0 for an acetate-oxidizing *Desulfobacter* sp.

Both APS reductase, an intracellular enzyme found in all SRB, and hydrogenase, an enzyme present in some SRB (hydrogenase-positive), can be extracted from liquids or solids, including corrosion products and sludge. In a procedure to quantify APS reductase, cells are lysed to release the enzyme, added to an antibody reagent, and exposed to a color-developing solution. In the presence of APS reductase, a blue color appears whose intensity and development rate is proportional to the amount of enzyme and roughly to the number of cells from which the enzyme was extracted. Similarly, hydrogenase activity may be measured by a procedure, where the enzyme is extracted from cells and exposed to hydrogen anaerobically (Boivin et al., 1990). The rationale for relating hydrogenase to MIC is that during corrosion in anaerobic environments, molecular hydrogen is produced at the cathode. Some, but not all, SRB are hydrogenase-positive, meaning that they possess the enzyme required to catalyze the formation of molecular hydrogen. In the assay, hydrogenase reacts with hydrogen and reduces an indicator dye in solution. The activity of hydrogenase is established by the development and intensification of a blue color, proportional to the rate of hydrogen uptake by the enzyme. The technique does not attempt to estimate specific numbers of SRB. Bryant et al. (1991) suggested that hydrogenase levels were better indicators of MIC than numbers of SRB. Mara and Williams (1972) reported that hydrogenase was more important when the environment contained low concentrations of ferrous ions and less important in the presence of sufficient ferrous ions to precipitate the sulfide produced by SRB. Other investigators found no relationship between the levels of hydrogenase enzyme and the rate or extent of corrosion (Jones–Meehan et al., 2003).

Cell Activity

Roszak and Colwell (1987) reviewed techniques commonly used to detect microbial activities in natural environments, including transformations of radiolabeled metabolic precursors. Phelps et al. (1991) and Mittelman et al. (1990) used uptake or transformation of ^{14}C-labeled metabolic precursors to examine activities of sessile bacteria in natural environments and in laboratory models. Phelps et al. (1991) used a variety of ^{14}C-labeled compounds to quantify catabolic and anabolic bacterial activities associated with corrosion tubercles in steel natural gas transmission pipelines. They demonstrated that organic acid was produced from hydrogen and carbon dioxide in natural gas by acetogenic bacteria, and that acidification could lead to enhanced corrosion of the steel. Mittelman et al. (1990) used measurement of lipid biosynthesis from ^{14}C-acetate, in conjunction with measurements of microbial biomass and extracellular polymer, to study the effects of differential fluid shear on physiology and metabolism of *Alteromonas* (formerly *Pseudomonas*) *atlantica*. Increasing shear force increased the rate of total lipid biosynthesis, but decreased per cell biosynthesis. Increasing fluid shear also increased cellular biomass and greatly increased the ratio of extracellular polymer to cellular protein. Maxwell (1986) developed a radio-respirometric technique for measuring SRB activity on metal surfaces that involved two distinct steps: incubation of the sample with ^{35}S-sulfate, and trapping the released sulfide.

Techniques for analyzing microbial metabolic activity at localized sites have also been developed. Franklin et al. (1992) incubated microbial biofilms with ^{14}C-metabolic precursors and autoradiographed the biofilms to localize biosynthetic activity on corroding metal surfaces. The localized uptake of labeled compounds was related to localized electrochemical activities associated with corrosion reactions.

Reporter genes can signal when the activity of a specific metabolic pathway is induced. King et al. (1990) engineered the incorporation of a promotorless cassette of *lux* genes into specific operons of *Pseudomonas* to induce bioluminescence during degradation of naphthalene. Using reporter genes, Marshall (1994) demonstrated that bacteria immobilized on surfaces exhibit physiological properties not found in the same organisms in the aqueous phase. Some genes were turned on at solid surfaces despite not being expressed in liquid or on solid media. Other genes were turned off on surfaces. They further demonstrated gene transfers within biofilms even in the absence of imposed selection pressure.

Genetic Techniques

Genetic techniques using ribosomal RNA (rRNA) or their genes (rDNA) have been used to identify and quantify microbial populations in natural environments (Stahl et al., 1984, 1988; Amann et al., 1992). These techniques involve amplification of 16S rRNA gene sequences by polymerase chain reaction (PCR) amplification of the extracted and purified nucleic acids. The PCR products can be evaluated using community-fingerprinting techniques such as denaturing gradient gel electrophoresis (DGGE). Each DGGE band is representative of a specific bacterial population and the number of distinctive bands is indicative of microbial diversity. The PCR products can also be sequenced and the sequences compared with those in the databases, which allows for identification of the species within an environmental sample.

Horn et al. (2003) identified the constituents of the microbial community within a proposed nuclear waste repository using two techniques: (1) isolation of DNA from growth culture and subsequent identification by 16S rDNA genes, and (2) isolation of DNA directly from environmental samples followed by subsequent identification of the amplified 16S rDNA genes (Table 3-1 and Figure 3-1). Comparison of data from these two techniques demonstrates that culture-dependent approaches underestimate the complexity of microbial communities. Zhu et al. (2003, 2004) used genetic techniques to characterize the types and abundance of bacterial species in gas pipeline samples and made similar observations. Another example of a genetic technique is fluorescent in situ hybridization (FISH), which uses specific fluorescent

TABLE 3-1   Organisms Isolated After Growth in Various Yucca Mountain Simulated
Groundwaters and 16S rDNA Sequence Divergence from Reference Organisms

	Growth medium from which organism is isolated	
	1×J13 Synthetic with glucose	1×J13 Synthetic without glucose
Organism[a]	% divergence from database[b]	
Ralstonia pickettii		
Ralstonia eutrophus	4.5/MS	6.9/MS
Burkholderia cepacia		
Blastobacter natatorius		
Sphingomonas paucimobilis		2.0/GB
Methylobacterium mesophilicum		
Caulobacter subvibroidies		
Uncultured bacterium oxSCC-6[c]	4.0/GB	
Pseudomonas (Janth) *mephitica*	6.06/MS	
Microbacterium barkeri	4.55/MS	
Microbacterium keratanolyticum	4.15/MS	4.15/MS
Microbacterium chocolatum	5.16/MS	
Arthrobacter sp. SMCC G964[d]	0.0/GB	
Pseudomonas stutzeri	3.93/MS	1.0/GB
Afipia genosp. 14		4.0/GB

Source: Horn et al. (2003, © NACE International).
[a]Closest relative in 16S rDNA sequence comparisons to three separate databases (i.e., MS, GB, and RDP,
below).
[b]MS, MicroSeq database; GB, GenBank database; RDP, Ribosomal database project.
[c]Lüdemann et al. (2000).
[d]Van Waasbergen et al. (2000).

dye-labeled oligonucleotide probes to selectively identify and visualize SRB both in
established and developing multispecies biofilms (Amann et al., 1992).

Microscopy

Light Microscopy

Using light microscopy and proper staining, investigators (Chamberlain et al., 1995,
1988) have demonstrated a relationship between an unusual variety of copper pitting
corrosion and gelatinous, polysaccharide-containing biofilms.

Epifluorescence Microscopy

Immunofluorescence techniques have been developed for the identification of specific bacteria in biofilms (Howgrave–Graham and Steyn, 1988; Zambon et al., 1984).

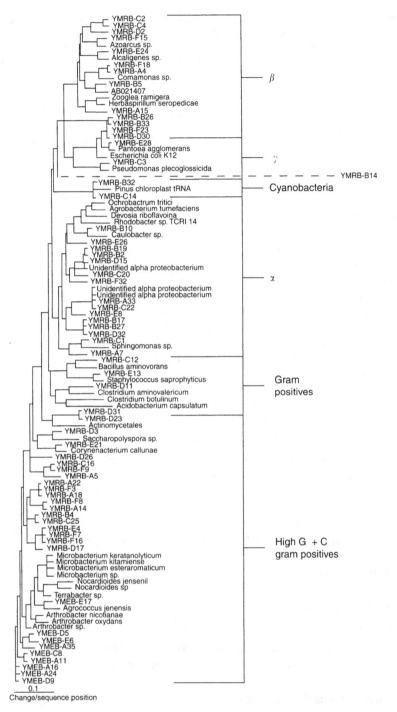

FIGURE 3-1 Phylogenetic tree of Yucca Mountain bacterial community as identified by 16S rDNA analysis of DNA extracted from Yucca Mountain rock. (Horn et al., 2003. © NACE International.)

Epifluorescence cell-surface antibody (ECSA) methods for detecting SRB are based on the binding between SRB-specific antibodies and SRB cells and subsequent detection of SRB-specific antibodies with a secondary antibody through two approaches. First, the secondary antibodies can be linked to a fluorochrome, which enables bacterial cells marked with the secondary antibody to be viewed with an epifluorescence microscope. Second, the secondary antibodies can be conjugated with an enzyme (alkaline phosphatase) that can then be reacted with a colorless substrate to produce a visible color, proportional to the quantity of SRB present. Detection limits for the field test are 10,000 SRB mm^{-2} filter area. The color reagent used for field tests is unstable at room temperature and tends to bind nonspecifically with antibodies adsorbed directly at active sites on the filter, creating a false-positive, which may interfere with the detection of SRB at levels below 10,000 cells mm^{-2}. Antigenic structures of marine and terrestrial strains are distinctly different and therefore antibodies to either strain did not react with the other.

Confocal Laser Scanning Microscopy

Confocal laser scanning microscopy (CLSM) permits one to create three-dimensional images, determine surface contour in minute detail, and accurately measure critical dimensions by mechanically scanning the object with laser light. A sharply focused image of a single horizontal plane within a specimen is formed while light from out-of-focus areas is repressed from view. The process is repeated at precise intervals on horizontal planes and visual data from all images compiled to create a single, multidimensional view of the subject. Geesey et al. (1994) used CLSM to produce three-dimensional images of bacteria within scratches, milling lines, and grain boundaries.

Atomic Force Microscopy

Atomic force microscopy (AFM) uses a microprobe mounted on a flexible cantilever to detect surface topography by scanning at a subnanometer scale. Repulsion by electrons overlapping at the tip of the microprobe causes deflections of the cantilever that can be detected by a laser beam. The signal is read by a feedback loop to maintain a constant tip displacement by varying voltage to a piezoelectric control. Variations in the voltage mimic the topography of the sample and, together with the movement of the microprobe in the horizontal plane, are converted into an image. Telegdi et al. (1998) imaged microorganisms associated with corrosion on several substrata using AFM.

Electron Microscopy

Much of the conclusions about biofilm development, composition, distribution, and relationship to substratum/corrosion products have been derived from traditional scanning electron microscopy (SEM) and transmission electron microscopy (TEM). Scanning electron microscopy has been used to image SRB from corrosion products on alloy 904L (UNS N08904) (Scott and Davies, 1989), microorganisms in corrod-ing gas pipelines (Zhu et al., 2003), and iron-oxidizing *Gallionella* in water distri-bution systems (Ridgeway and Olsen, 1981). Transmission electron microscopy has been used to demonstrate that bacteria are intimately associated with sulfide miner-als and that on copper-containing surfaces the bacteria are found between alternate layers of corrosion products and attached to base metal (Blunn, 1986).

In traditional SEM, nonconducting samples, including biofilms associated with corrosion products, must be dehydrated, fixed, and coated with a conductive film of metal before the specimen can be viewed. Traditional TEM methods for imaging biofilms require fixation of biological material, embedded in a resin and thin sectioning to achieve a section that can transmit an electron beam. Environmental electron microscopy includes both scanning (ESEM) and transmission (ETEM) techniques for the examination of biological materials with a minimum of manipulation, that is, fix-ation and dehydration. Little et al. (1991) used ESEM to study marine biofilms on 304 stainless-steel (UNS S30400) and 90/10 copper-nickel (UNS C70600) surfaces. They observed a gelatinous layer in which bacteria and microalgae were embedded. Traditional SEM images of the same areas demonstrated a loss of cellular and extra-cellular material (Figure 3-2a, b). Ray and Little (2003) and Little et al. (1998) used ESEM to demonstrate sulfide-encrusted SRB in corrosion layers on pure copper (UNS C11000) and C70600 (Figure 3-3a, b) and iron-depositing bacteria in tubercles on galvanized steel pipe (undesignated) (Figure 3-4a–c). Little et al. (2001) used ETEM to image *P. putida* on corroding iron filings and to demonstrate that the organ-isms were not directly in contact with the metal. Instead, the cells were attached to the substratum with extracellular material (Figure 3-5). Design and operation of the ESEM and ETEM have been described elsewhere (Ray and Little, 2003).

There are fundamental problems in attempting to diagnose MIC by establishing a spatial relationship between number and types of microorganisms in the bulk medium or those associated with corrosion products using any of the techniques described previously. Zintel et al. (2003) established that there were no relationships

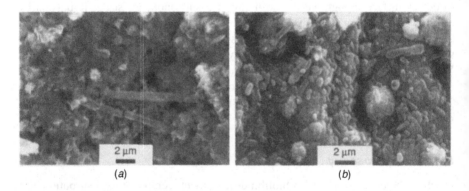

(a)   (b)

FIGURE 3-2a, b   (a) ESEM image of bacteria within sulfide corrosion layers on 90/10 copper-nickel (UNS C70600) foil (marker = 2μm). (b) SEM image of the same area after traditional fixation, dehydra-tion and critical-point drying (marker = 2μm). (Little et al., 1991.)

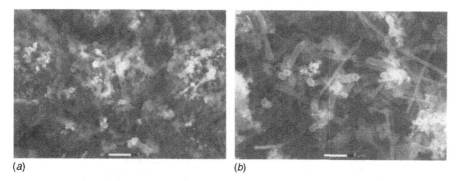

(a)   (b)

FIGURE 3-3a, b ESEM images of bacteria within corrosion layers on UNS C11000 copper foils (markers = 5μm). (Ray and Little, 2003.)

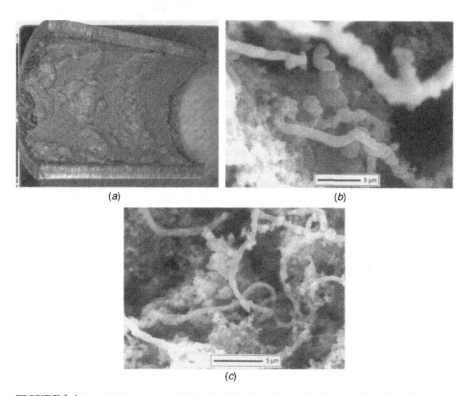

(a)   (b)

(c)

FIGURE 3-4a–c (a) Tubercles associated with pitting in galvanized steel (undesignated) pipe from water distribution system (2x), (b, c) ESEM views of twisted filaments observed in tubercles. (Ray and Little, 2003.)

between the presence, type, or levels of planktonic or sessile bacteria and the occurrence of pits. Because microorganisms are ubiquitous, the presence of bacteria or other microorganisms does not necessarily indicate a causal relationship with corrosion. In fact, microorganisms can nearly always be cultured or imaged from natural

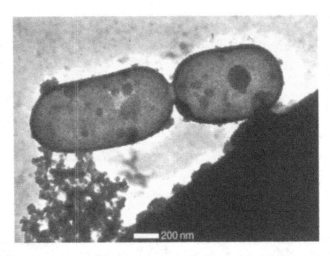

FIGURE 3-5   TEM micrograph of hydrated *Pseudomonas putida* after removal of excess moisture by circulation of air through the environmental cell. (Little et al., 2001.)

environments. Little et al. (1996) reported that electrochemical polarization could influence the number and types of bacteria associated with the surface. Artificial crevices created in S30400 in abiotic seawater were associated with large numbers of bacteria after 5-day exposures to natural seawater. Bacteria did not cause the crevice, instead bacteria were attracted to this anodic site. Several other investigators have made similar observations. For example, de Sanchez and Schiffrin (1996) demonstrated that Cu (II) and titanium ions were strong attractants for *Pseudomonas*. Detection or demonstration of bacteria associated with corrosion products or corrosion sites is not an independent diagnostic for MIC.

PIT MORPHOLOGY

Pope (1990) completed a study of gas pipelines to determine the relationship between the extent of MIC and the levels/activities of SRB. He concluded that there was no relationship. Instead, he found large numbers of APB and organic acids, particularly lactic acid, and identified the following metallurgical features on an undesignated carbon steel alloy:

- Large craters from 5 to 8 cm or greater in diameter surrounded by uncorroded metal (Figure 3-6)

- Cup-type hemispherical pits on the pipe surface or in the craters (Figure 3-7)

- Striations or contour lines in the pits or craters running parallel to longitudinal pipe axis (rolling direction) (Figure 3-8)

- Tunnels at the ends of the craters also running parallel to the longitudinal axis of the pipe (Figure 3-9).

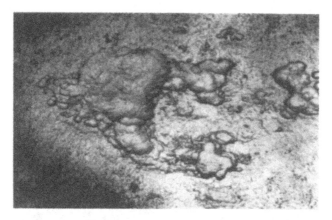

FIGURE 3-6   Closeup of sandblasted surface showing pits attributed to MIC. (Reprinted from Pope, 1990, with permission of the Gas Research Institute.)

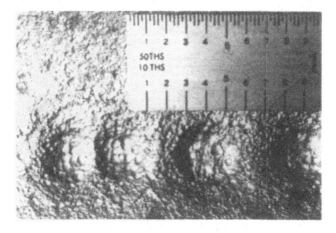

FIGURE 3-7   Cup-type scooped-out hemispherical pits on flat surfaces with craters in pits. (Reprinted from Pope, 1990, with permission of the Gas Research Institute.)

Pope (1990) reported that these metallurgical features were "fairly definitive for MIC." However, the author did not advocate diagnosis of MIC based solely on pit morphology. Subsequent research has demonstrated that these features can be produced by abiotic reactions (Eckert et al., 2003a) and cannot be used to independently diagnose MIC.

Other investigators described "ink bottle"-shaped bulbous pits in 300 series stainless steel that were supposed to be diagnostic of MIC (Figure 3-10). Borenstein and Lindsay (1994, 1988) reported that dendritic corrosion attack at welds was "characteristic of MIC." Hoffman (1993) suggested that pit morphology was a "metallurgical fingerprint... definitive proof of the presence of MIC." Chung and Thomas (1999) compared MIC pit morphology with non-MIC chloride-induced pitting in

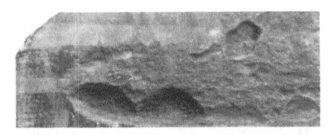

FIGURE 3-8 Corrosion pits with striations. (Reprinted from Pope, 1990, with permission of the Gas Research Institute.)

FIGURE 3-9 Closeup view of tunnels (100×). (Reprinted from Pope, 1990, with permission of the Gas Research Institute.)

S30400, 304L (UNS S30403), and 308 (UNS S30800) stainless-steel base metals and welds. A faceted appearance was common to both types of pits in S30400 and S30403 base metal (Figure 3-11a, b). Facets were located in the dendritic skeletons in MIC and non-MIC cavities of 308 (UNS S30800) weld metal. They concluded that there were no unique morphological characteristics for MIC pits in these materials.

The problem that has resulted from the assumption that pits can be independently interpreted as MIC is that MIC is often misdiagnosed. For example, Welz and Tverberg (1998) reported that leaks at welds in 316L stainless-steel (UNS S31603) hot water system in a brewery after 6 weeks in operation were due to MIC. The original diagnosis was based on the circumstantial evidence of attack at welds and the pitting morphology-scalloped pits within pits. However, after a thorough investigation MIC was dismissed. There were no bacteria associated with the corrosion sites, and the deposits were too uniform to have been produced by bacteria. The hemispherical pits had been produced when CO_2 was liberated and low pH bubbles nucleated at surface discontinuities.

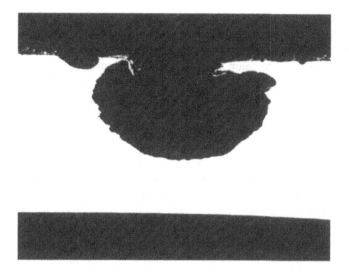

FIGURE 3-10   An illustration of an "ink bottle" type pit noted in many cases of MIC, especially in stainless steels containing less than 6% molybdenum.

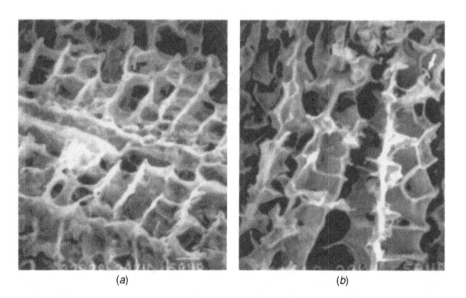

(a)   (b)

FIGURE 3-11 a, b   (a) SEM micrograph of dendritic skeletons in MIC cavities in E308 stainless-steel weld (1000×). (b) Micrograph of the non-MIC chloride-induced corrosion pits in 308 (UNS S30800) weld root (300×). (Chung and Thomas, 1999. © NACE International.)

More recently, several investigators have demonstrated that the initial stages of pit formation, due to certain types of bacteria, do have unique characteristics. Geiser et al. (2001) found that pits formed in S31603 stainless steel attributed to the manganese-oxidizing bacterium *Leptothrix discophora* had different morphologies than pits initiated by anodic polarization. Pits initiated by these organisms in sodium chloride solution were approximately 10 times as long as they were wide (Figure 3-12a, b). Pits produced by microorganisms were much smaller than and not nearly as deep as pits produced in the same solution by electrochemical means. Pits had almost identical sizes and aspect ratios as the sizes and aspect ratios of the manganese-oxidizing bacteria. The similarity between the dimensions of bacterial cells attached to the surface and the dimensions of corrosion pits indicates a possibility that the pits were initiated at the sites where the microbes were attached. Eckert et al. (2003b) used pipeline steel (American Petroleum Designation 5L) to demonstrate micromorphological characteristics that could be used to identify MIC initiation. Coupons were installed at various points in a pipeline system and examined by SEM at 1000 and 2000×. They demonstrated that pit initiation and bacterial colonization were correlated and that pit locations physically matched the locations of cells. Telegdi et al. (1998) used AFM to image biofilm formation, extracellular polymer production, and subsequent corrosion. Pits produced by *Thiobacillus intermedius* had the same shape as the bacteria. None of the investigators claims that these unique features can be detected with the unaided eye or that these features will be preserved as pits grow, propagate, and merge.

CHEMICAL TESTING

Analyses for corrosion product chemistry can range from simple field tests to mineralogy and isotope fractionation. Field tests for solids and corrosion products typically include pH and a qualitative analysis for the presence of sulfides and carbonates. A drop of dilute hydrochloric acid placed on a small portion of the corrosion product will indicate the presence of carbonates if noticeable bubbling occurs. When the acid-treated corrosion product is exposed to lead acetate test paper, a color

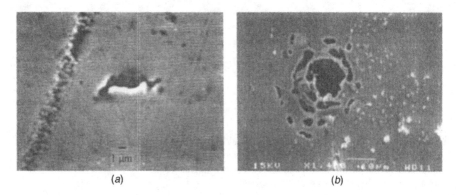

(a)   (b)

FIGURE 3-12a, b (a) SEM image showing heavy line on left indicating square etching by iron milling. The indention in the center was detected after the coupon was microbially ennobled, and was not present before microbial colonization. (b) SEM image of a typical pit initiated in 316L stainless steel (UNS S31603) using anodic polarization. (Geiser et al., 2001. © NACE International.)

change from white to brown will occur or will produce the smell of rotten eggs if sulfide is present.

Elemental Composition

Elements in corrosion deposits can provide information as to the cause of corrosion. Energy-dispersive x-ray microanalysis (EDS) coupled with an electron microscope can be used to determine the elemental composition of corrosion deposits. Because all living organisms contain ATP, a phosphorus peak in an EDS spectrum can be related to cells associated with the corrosion products. Other sources of phosphorus, for example, phosphate water treatments, must be eliminated. The activities of SRB and manganese-oxidizing bacteria produce surface-bound sulfur and manganese, respectively. Chloride is typically found in crevices and pits and cannot be directly related to MIC.

There are several limitations for EDS surface chemical analyses. Samples for EDS cannot be evaluated after heavy metal coating, so EDS spectra must be collected prior to heavy-metal coating. It is difficult or impossible to match spectra with exact locations on images. This is not a problem with ESEM because nonconducting samples can be imaged directly, meaning that EDS spectra can be collected of an area that is being imaged by ESEM. Little et al. (1991) documented the changes in surface chemistry as a result of solvent extraction of water, a requirement for SEM (Table 3-2). Other shortcomings of SEM/EDS include peak overlap. Peaks for sulfur overlap with peaks for molybdenum, and the characteristic peak for manganese coincides with the secondary peak for chromium. Wavelength-dispersive spectroscopy can be used to resolve overlapping EDS peaks. Peak heights cannot be used to determine the concentration of

TABLE 3-2   Weight Percent of Elements Found on Commercially Pure Copper Surfaces after Exposure to Estuarine Water for 4 Months and Sequential Treatment with Acetone and Xylene for Traditional SEM Imaging

Element	Base metal	After exposure to estuarine water	After acetone	After xylene
Al		9.49	1.22	0.74
Si		21.38	1.89	1.27
Cl		0.93	15.90	15.93
Cu	99.9	59.62	80.99	82.06
Mg		1.96	0	0
P		0.98	0	0
S		0.95	0	0
Ca		0.49	0	0
K		0.67	0	0
Fe		3.52	0	0

Source: Little et al. (1991).

elements. It is also impossible to determine the form of an element with EDS. For example, a high sulfur peak may indicate sulfide, sulfate, or elemental sulfur.

Mineralogical Fingerprints

McNeil et al. (1991) used mineralogical data determined by x-ray crystallography, thermodynamic stability diagrams (Pourbaix), and the simplexity principle for precipitation reactions to evaluate corrosion product mineralogy. They concluded that many sulfides under near-surface natural environmental conditions could only be produced by microbiological action on specific precursor metals. They reported that djurleite, spionkopite, and the high-temperature polymorph of chalcocite were mineralogical fingerprints of SRB-induced corrosion of copper–nickel alloys. They also reported that the stability or tenacity of sulfide corrosion products determined their influence on corrosion (detailed in Chapter 2).

Jack et al. (1995) maintained that the mineralogy of corrosion products on gas transmission pipelines (undesignated) could provide insights into the conditions under which the corrosion took place. For example, under anaerobic conditions in the absence of SRB, an iron (II) carbonate was identified in water trapped under defective coatings. Introduction of air caused a rapid discoloration of the white corrosion product to orange iron (III) oxides. In the presence of SRB, indicator minerals are siderite and iron (II) sulfide in a ratio of 3:1 or more (Figure 3-13). Mackinawite, the first formed crystalline sulfide, converts into gregite in a time- and pH-dependent manner. Conversion is only rapid at pH below 6. At pH 6, the conversion is complete in about 3 months. At pH 7, it takes 6 months and at pH 8 mackinawite persists along with gregite, and pyrrhotite may form after 9 months. At aerobic corrosion sites, the minerals are iron (III) oxides, magnetite, hematite, lepidocrocite, and goethite (Figure 3-14).

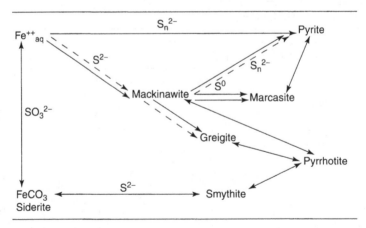

FIGURE 3-13   Transformations of Fe(II) sulfides formed at pipeline corrosion sites (dashes, biological processes; solid lines, abiological processes). (Jack et al., 1995. © NACE International.)

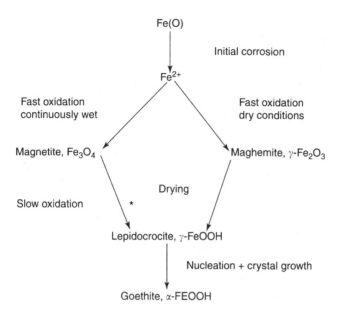

* Hematite may form from magnetite as an intermediate.

FIGURE 3-14   Transformation of Fe(III) oxides formed at pipeline corrosion sites. (Jack et al., 1995. © NACE International.)

Isotope Fractionation

The stable isotopes of sulfur (^{32}S and ^{34}S), naturally present in any sulfate source, are selectively metabolized during sulfate reduction by SRB and the resulting sulfide is enriched in ^{32}S (Chambers and Trudinger, 1979). ^{34}S accumulates in the starting sulfate as ^{32}S is removed and becomes concentrated in the sulfide. Little et al. (1993) demonstrated sulfur isotope fractionation in sulfide corrosion deposits resulting from activities of SRB within biofilms on C70600 copper-nickel surfaces. ^{32}S accumulated in sulfide-rich corrosion products, while ^{34}S was concentrated in the residual sulfate in the culture medium. Accumulation of the lighter isotope was related to surface derivatization or corrosion as measured by weight loss. Use of this technique to identify SRB-related corrosion requires sophisticated laboratory procedures.

SUMMARY

The following are required for an accurate diagnosis of MIC: (1) a sample of the corrosion product or affected surface that has not been altered by collection or storage, (2) identification of a corrosion mechanism, (3) identification of microorganisms capable of growth and maintenance of the corrosion mechanism in the particular environment, and (4) demonstration of an association of the microorganisms with the observed corrosion. Three types of evidence are used to diagnosis MIC: metallurgical, chemical, and biological. The objective is to have three independent types of measurements that are consistent with a mechanism for MIC.

It is essential in diagnosing MIC to demonstrate a spatial relationship between the causative microorganisms and the corrosion phenomena. However, that relationship cannot be independently interpreted as MIC. Pitting due to MIC can initiate as small pits that have the same size and characteristics of the causative organisms. These features are not obvious to the unaided eye and are most often observed with an electron or atomic force microscope. MIC does not produce a macroscopic unique metallographic feature. Metallurgical features previously thought to be unique to MIC, for example, hemispherical pits in 300 series stainless steel localized at welds or tunneling in carbon steel, are consistent with some mechanisms for MIC, but cannot be interpreted independently. Bacteria produce corrosion products that cannot be produced abiotically in near-surface environments, resulting in isotope fractionation and mineralogical fingerprints. The result is corrosion where none could be anticipated based on the composition of the bulk medium, for example, low-chloride waters, and rates that are exceptionally fast. Development of sophisticated genetic and imaging techniques has made it possible to more accurately characterize microorganisms and their spatial relationships to corrosion products and localized corrosion.

REFERENCES

Amann RI, Stromley J, Devereux R, Key R, Stahl DA (1992). Molecular and microscopic identification of sulfate-reducing bacteria in multispecies biofilms. *Appl. Environ. Microbiol.*, **58**(2): 614.

American Petroleum Institute (API), 1965. API recommended practice for biological analysis of subsurface injection waters – RP-38. New York.

ASTM (1977). Use of ATP extraction in oil field waters. *STP*, **641**: 79. Philadelphia.

Blunn G (1986). Biological fouling of copper and copper alloys. *Biodeterioration*, **6**: 567–575.

Boivin J, Laishley EJ, Bryant RD, Costerton JW (1990). The influence of enzyme systems on MIC. *CORROSION/90*, Paper No. 128. Houston, TX: NACE International, p. 8.

Borenstein SW, Lindsay PB (1988). Microbiologically influenced corrosion failure analyses. *Mater. Perform.*, **27**(3): 51.

Borenstein SW, Lindsay PB (1994). MIC failure analysis. *Mater. Perform.*, **33**(4): 43.

Bryant W, Jansen W, Boivin J, Laishley EJ, Costerton JW (1991). Effect of hydrogenase and mixed sulfate-reducing bacterial populations on the corrosion of steel. *Appl. Environ. Microbiol.*, **57**(10): 2804.

Chamberlain AHL, Angell P, Campbell HS (1988). Staining procedures for charaterizing biofilms in corrosion investigations. *Br. Corros. J.*, **23**: 197.

Chamberlain AHL, Fischer WR, Hinze U, Paradies HH, Sequeira CAC, Siedlarek H, Thies M, Wagner D, Wardell JN (1995). An interdisciplinary approach for microbiologically influenced corrosion of copper. *Microbial Corrosion, Proceedings of the 3rd International EFC Workshop*, Vol. 15. London: The Institute of Materials, p. 3.

Chambers LA, Trudinger PA (1979). Microbiological fractionation of stable sulfur isotopes: a review and critique. *Geomicrobiol. J.*, **1**: 249.

Chung Y, Thomas LK (1999). Comparison of MIC pit morphology with non-MIC chloride induced pits in types 304/304L/E308 stainless steel base metal/welds. *CORROSION/99*, Paper No. 159. Houston, TX: NACE International, p. 23.

Cowan JK (2005). Rapid enumeration of sulfate-reducing bacteria. *CORROSION/2005*, Paper No. 05485. Houston, TX: NACE International, p. 16.

Dowling NJE, Nichols PD, White DC (1988). Phospholipid fatty acid and infrared spectroscopic analysis of a sulphate-reducing consortium. *FEMS Microbiol. Ecol.*, **53**: 325.

Eckert R (2003a). *Field Guide for Investigating Internal Corrosion of Pipelines*. Houston, TX: NACE International, p. 21.

Eckert RB, Aldrich HC, Edwards CA, Cookingham BA (2003b). Microscopic differentiation of internal corrosion initiation mechanisms in natural gas pipeline systems. *CORROSION/2003*, Paper No. 03544. Houston, TX: NACE International, p. 13.

Franklin MJ, White DC (1991). Biocorrosion, current opinion. *Biotechnology*, **2**: 450.

Franklin MJ, White DC, Isaacs HS (1992). A study of carbon steel corrosion inhibition by phosphate ions and by organic buffer, using a scanning vibrating electrode. *Corrosion Sci.*, **33**: 251.

Geesey GG, Lewandowski Z, Flemming HC (1994). *Biofouling and Biocorrosion in Industrial Water Systems*, Boca Raton, FL: CRC Press, cover.

Geiser M, Avci R, Lewandowski Z (2001). Pit initiation on 316L stainless steel in the presence of bacteria *Leptothrix discophora*. *CORROSION/2001*, Paper No. 01257. Houston, TX: NACE International, p. 10.

Giovannoni SJ, Britschgi TB, Moyer CL, Field KG (1990). Genetic diversity in Sargasso sea bacterioplankton. *Nature*, **344**: 60.

Hoffman RA (1993). Case histories of microbiologically influenced corrosion in building and power generation systems. *CORROSION/93*, Paper No. 317. Houston, TX: NACE International, p. 19.

Hogan JJ (1990). A rapid, non-radioactive DNA probe for detection of SRBs. *Institute of Gas Technology Symposium on Gas, Oil, Coal, and Environmental Biotechnology*. Illinois: IGT.

Horn J, Carrillo C, Dias V (2003). Comparison of the microbial community composition at Yucca mountain and laboratory test nuclear repository environments. *CORROSION/2003*, Paper No. 03556. Houston, TX: NACE International, p. 12.

Howgrave-Graham AR, Steyn PL (1988). Application of the fluorescent-antibody technique for the detection of *Sphaerotilus natans* in activated sludge. *Appl. Environ. Microbiol.*, **54**(3): 799.

Jack TR, Wilmott MJ, Sutherby RL (1995). Indicator minerals formed during external corrosion of line pipe. *Mater. Perform.*, **34**: 19.

Jhobalia C, Hu A, Gu T, Nesic S (2005). Biochemical engineering approaches to MIC. *CORROSION/2005*, Paper No. 05500. Houston, TX: NACE International, p. 12.

Jones-Meehan J, Cofield JW, Little B, Ray R, Wagner P, McNeil M, McKay J (2003). Microbiologically influenced corrosion of steels: weight loss measurements, ESEM/EDS and XRD analyses. *Int. Conf. of MIC*, Paper No. 29. Houston, TX: NACE International, p. 12.

Kaeberlein T, Lewis K, Epstein SS (2002). Isolating "uncultivable" microorganisms in pure culture in a simulated natural environment. *Science*, **226**: 1127.

King JMH, DiGrazia PM, Applegate B, Buriage R, Sanseverino J, Dunbar P, Larimer F, Saylor GS (1990). Rapid, sensitive bioluminescent reporter technology for napthalene exposure and biodegradation. *Science*, **249**: 778.

Little B, Wagner P, Ray R, McNeil M, Jones-Meehan J (1993). Indicators of microbiologically influenced corrosion in copper alloys. *Oebalia* (XIX, Suppl.): 287.

Little B, Wagner P, Ray R, Pope R, Scheetz R (1991). Biofilms: an ESEM evaluation of artifacts introduced during SEM preparation. *J. Ind. Microbiol.*, **8**: 213.

Little BJ, Pope RK, Daulton T, Ray RI (2001). Application of transmission electron microscopy to microbiologically influenced corrosion. *CORROSION/2001*, Paper No. 01266. Houston, TX: NACE International.

Little BJ, Wagner PA, Hart KR, Ray RI (1996). Spatial relationships between bacteria and localized corrosion. *CORROSION/96*, Paper No. 278. Houston, TX: NACE International, p. 8.

Little BJ, Wagner PA, Lewandowski Z (1998) The role of biomineralization in microbiologically influenced corrosion. *CORROSION/98*, Paper No. 294. Houston, TX: NACE International, p. 18.

Lloyd B (1937). Bacteria stored in seawater. *J. Roy Tech Coll. (Glasgow)*, **4**: 173.

Lüdemann H, Arth I, Liesack W (2000). Spatial changes in the bacterial community structure along a vertical oxygen gradient in flooded paddy soil cores. *Appl. Environ. Microbiol.*, **66**: 754–762.

Mara DD, Williams DJA (1972). Influence of the microstructure of ferrous metals on the rate of microbial corrosion. *Br. Corros. J.*, **7**(5): 139.

Marshall KC (1994). Analysis of bacterial behavior during biofouling of surfaces. In: Geesey G.G., Lewandowski Z, Flemming HC (eds), *Biofouling and Biocorrosion in Industrial Water Systems*. Boca Raton, FL: Lewis Publishers, p. 15.

Maxwell S (1986). Assessment of sulfide corrosion risks in offshore systems by biological monitoring. *SPE Prod. Eng.*, Vol. 1: p. 363.

McNeil MB, Jones JM, Little BJ (1991). Mineralogical fingerprints for corrosion processes induced by sulfate-reducing bacteria. *CORROSION/91*, Paper No. 580. Houston, TX: NACE International, p. 16.

Mittelman MW, Nivens DE, Low C, White DC (1990). Differential adhesion, activity, and carbohydrate: protein ratios of *Pseudomonas Atlantica* monocultures attaching to stainless steel in a linear shear gradient. *Microbial Ecol.*, **19**: 269.

Phelps TJ, Schram RM, Ringelberg D, Dowling NJ, White DC (1991). Anaerobic microbial activities including hydrogen-mediated acetogenesis within natural gas transmission lines. *Biofouling*, **3**: 265.

Pope DH (1986). Discussion of methods for the detection of microorganisms involved in microbiologically influenced corrosion. *Biologically Induced Corrosion, Proceedings of the International Conference on Biologically Induced Corrosion*. Houston, TX: NACE International, p. 275.

Pope DH (1990). Microbiologically influenced corrosion (MIC): methods of detection in the field. *GRI Field Guide*. Illinois: Gas Research Institute.

Postgate JR (1979). *The Sulphate-Reducing Bacteria*. Cambridge, UK: Cambridge University Press, p. 26.

Ray R, Little B (2003). Environmental electron microscopy applied to biofilms. In: Lens P, Moran AP, Mahony T, Stoodley P, O'Flaherty V (eds) *Biofilms in Medicine, Industry and Environmental Biotechnology*. London: IWA Publishing, p. 331.

Ridgeway HF, Olson BH (1981). Scanning electron microscope evidence for bacterial colonization of a drinking water distribution system. *Appl. Environ. Microbiol.*, **41**(1): 274.

Romero JM, Velázquez E, Garcia-Villalobos JL, Amaya M, Le Borgne S (2005). Genetic monitoring of bacterial populations in a seawater injection system. Identification of biocide resistant bacteria and study of their corrosive effect. *CORROSION/2005*, Paper No. 05483. Houston, TX: NACE International, p. 9.

Roszak DB, Colwell RR (1987). Survival strategies of bacteria in the natural environment. *Microbiology*, **51**(3): 365.

de Sánchez SR, Schiffrin DJ (1996). Bacterial chemo-attractant properties of metal ions from dissolving electrode surfaces. *J. Elect. Chem.*, **403**: 39.

Scott PJB, Davies M (1989). MIC of alloy 904L. *Mater. Perform.*, **28**(5): 57.

Scott PJB, Davies M (1992). Survey of field kits for sulfate-reducing bacteria. *Mater. Perform.*, **31**(5): 64.

Stahl DA, Flesher B, Mansfield HR, Montgomery L (1988). The use of phylogenetically based hybridization probes for studies of ruminal microbial ecology. *Appl. Environ. Microbiol.*, **54**: 1079.

Stahl DA, Lane DJ, Olsen GL, Pace NR (1984). Analysis of hydrothermal vent-associated symbionts by ribosomal RNA sequences. *Science*, **224**: 409.

Tatnall RE, Stanton KM, Ebersole RC (1988). Methods of testing for the presence of sulfate-reducing bacteria. *CORROSION/88*, Paper No. 88. Houston, TX: NACE International, p. 34.

Telegdi J, Keresztes Z, Páalinkás G, Kálmán E, Sand W (1998). Microbially influenced corrosion visualized by atomic force microscopy. *Appl. Phys.* A **66**: S639.

Van Waasbergen LG, Balkwill DL, Crocker FH, Bjornstad BN, Miller RV (2000). Genetic diversity among *Arthrobacter* species collected across a heterogeneous series of terrestrial deep-subsurface sediments as determined on the basis of 16S rRNA and recA sequences. *Appl. Environ. Microbiol.*, **66**: 3454–3463.

Ward DM, Ferris MJ, Nold SC, Bateson MM (1998). A natural view of microbial biodiversity within hot spring cyanobacterial mat communities. *Microbiol. Molec. Biol. Rev.*, **62**(4): 1353.

Welz J, Tverberg J (1998). Case history: corrosion of a stainless steel hot water system in a brewery. *Mater. Perform.*, **37**: 66.

Zambon JJ, Huber PS, Meyer AE, Slots J, Fornalik MS, Baier RE (1984). In-situ identification of bacterial species in marine microfouling films by using an immunofluorescence technique. *Appl. Environ. Microbiol.*, **48**(6): 1214.

Zhu X, Ayala A, Modi H, Kilbane JJ II (2005). Application of quantitative, real-time PCR in monitoring microbiologically influenced corrosion (MIC) in gas pipelines. *CORROSION/2005*, Paper No. 05493. Houston, TX: NACE International, p. 20.

Zhu X, Lubeck J, Kilbane JJ II (2003). Characterization of microbial communities in gas industry pipelines. *Appl. Environ. Microbiol.*, **69**(9): 5354.

Zhu X, Lubeck J, Lowe K, Daram A, Kilbane JJ II (2004). Improved method for monitoring microbial communities in gas pipelines. *CORROSION/2004*, Paper No. 04592. Houston, TX: NACE International, p. 13.

Zintel TP, Kostuck DA, Cookingham BA (2003). Evaluation of chemical treatments in natural gas systems versus MIC and other forms of internal corrosion using carbon steel coupons. *CORROSION/2003*, Paper No. 03574. Houston, TX: NACE International, p. 6.

Zobell C, Anderson DQ (1936). Effect of volume on bacterial activity. *Biol. Bull.*, **71**: 324.

Chapter 4

Electrochemical Techniques Applied to Microbiologically Influenced Corrosion

INTRODUCTION

Electrochemical techniques used to study microbiologically influenced corrosion (MIC) include those in which no external signal is applied [e.g., measurement of redox potential ($E_{r/o}$) or corrosion potential (E_{corr}), and electrochemical noise analysis (ENA)], those in which only a small potential or current perturbation is applied [e.g., polarization resistance (R_p) and electrochemical impedance spectroscopy (EIS)], and those in which the potential is scanned over a wide range (e.g., anodic and cathodic polarization curves, pitting scans) (Little and Wagner, 2001). Committee G1 of the American Society for Testing Materials publishes standard practices and experimental approaches for the most frequently used electrochemical techniques. Applications for mechanistic studies of MIC and/or for monitoring corrosion are presented in the following sections.

TECHNIQUES REQUIRING NO EXTERNAL SIGNAL

Redox Potential

A prepassivated platinum electrode and an electrode of the metal of interest were used to follow biofilm development and its effects on the corrosion behavior of structural materials (Zhang et al., 1989). Since the potential of platinum changes in the positive direction when either the oxygen concentration is increased or pH is

Microbiologically Influenced Corrosion By Brenda J. Little and Jason S. Lee
Published 2007 by John Wiley & Sons, Inc.

decreased, an indicator was added to the solution to determine pH changes independently. Time dependence of the open-circuit potential (E_{corr}) of several stainless steels and titanium was compared to that of the prepassivated platinum electrode. Figure 4-1a shows the time dependence of E_{corr} for platinum in solutions with and without bacteria, while Figure 4-1b shows the corresponding data for 304L stainless steel (UNS S30403). The authors (Zhang et al., 1989) concluded that the data reflect bacterial growth, that is, changes in the local environment induced by metabolic activities of the bacteria, and the resulting corrosion of the metal.

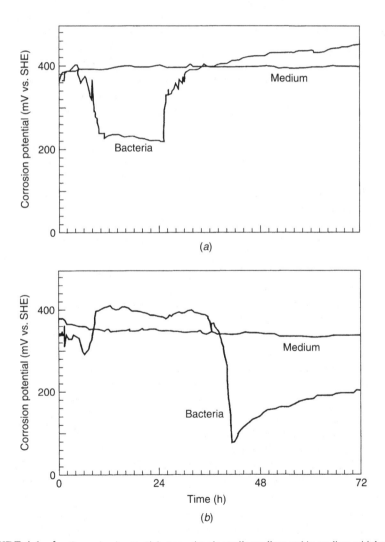

FIGURE 4-1a, b   Open circuit potential versus time in sterile medium and in medium with bacteria: (a) platinum, and (b) 304L stainless steel (UNS S30403). (Zhang et al., 1989. © NACE International.)

Open Circuit or Corrosion Potential, E_{corr}

E_{corr} measurements require a stable reference electrode — usually assumed to be unaffected by biofilm formation — and a high-impedance voltmeter. E_{corr} values are difficult to interpret, especially when related to MIC (Little and Wagner, 2001). Although ennoblement has been observed for metals exposed to both freshwater (rivers and estuaries) and natural seawater, the mechanisms may be different.

Results by Linhardt (1994, 1996), Renner (1996), Olesen et al. (2000a, b), Dickinson et al. (1996a), Dickinson and Lewandowski (1996), and Ruppel et al. (2001) present a specific mechanism for ennoblement of stainless steels in river water and estuaries involving microbial deposition of MnO_2. Linhardt (1994, 1996) found large amounts of manganese minerals (mainly $MnOOH$ and MnO_2) on severely pitted turbine runner blades (UNS J19540 and S41500) in a hydroelectric plant and suggested that pitting was due to biomineralized manganese oxides. Renner (1996) reported severe pitting around welds in 316 stainless-steel (UNS S31600) pipes used to pump cooling water from the Rhine River. He suggested that the failure mechanism involved iron- and manganese-oxidizing bacteria causing ennoblement and pitting.

Both Linhardt (1994) and Dickinson et al. (1996a) demonstrated that microbiologically deposited manganese oxide on 316L stainless-steel (UNS S31603) coupons in freshwater (Figure 2-16a–d) (Chapter 2) caused an increase in E_{corr} and increased cathodic current density at potentials above -200 mV versus saturated calomel electrode (SCE). For further details, see Chapter 2.

Dickinson et al. (1996a) measured E_{corr} (SCE) for S31603 coupons for up to 120 days exposure in fresh river water. E_{corr} was measured hourly by computer using a high-impedance ($>10^{11} \Omega$), multiplexed, analog-to-digital converter. Ennobled potentials >300 mV_{SCE} resulted from the formation of biofilms 2–40 µm thick and at least 3–5 percent coupon area coverage. E_{corr} measurements after removal of the biofilm from the ennobled metal surface and after subsequent reformation established that adherent fouling was necessary for initiation and retention of ennoblement.

Many investigators (Johnsen and Bardal, 1985; Mollica and Trevis, 1976; Scotto et al., 1985) have documented the ennoblement of passive alloys exposed in marine environments due to biofilm formation. The alloys tested include, but are not limited to, UNS S30400, S30403, S31254, S31600, S31603, S31703, S31803, S44635, S44660, S20910, S44735, N08028, N08367, N08904, N10276, NU984LN (non-UNS designation) and R50250. Crevice corrosion is the most problematic issue affecting the performance of stainless steels in seawater. The problem is exacerbated in warm natural seawater where biofilms form rapidly and shift E_{corr} in the noble direction toward the breakdown potential (E_b) for crevice corrosion initiation. Tests on passive alloys S30400 and S31600 indicated that crevice initiation times were reduced when natural marine biofilms were allowed to form on the boldly exposed external cathode surface of crevice corrosion samples (Zhang and Dexter, 1995). Even though there was variability in the data, the authors concluded that ennoblement of E_{corr} in the presence of natural biofilms usually caused the alloy to reach its E_b for crevice initiation faster than when the test was repeated under control conditions without the biofilm.

Most published E_{corr} data for stainless steels in natural seawater show rapid ennoblement during the first days of exposure. Figure 2-1 (Chapter 2) shows the potential–time curves reported by Johnsen and Bardal (1985) for several stainless steels (UNS S31603, N08904, N08028, S31254, S44635 and NU984LN) in flowing (0.5 m s^{-1}) natural seawater. E_{corr} changed from -200 to -250 mV$_{SCE}$ versus SCE at the beginning of the test to -50 mV$_{SCE}$ for two of the alloys and to $+200$ mV$_{SCE}$ for the remaining four alloys after 28 days. Explanations for microbially influenced ennoblement in seawater vary among investigators (Mollica and Trevis, 1976; Johnsen and Bardal, 1985; Scotto et al., 1985; Chandrasekaran and Dexter, 1993; Eashwar et al., 1992, 1995a,b).

Electrochemical Noise Analysis (ENA)

Electrochemical noise (EN) data can be obtained as fluctuations of E_{corr}, as fluctuations of potential (E) at an applied current (I), or as fluctuations of I at an applied E. No external signals that may influence biofilm properties are applied. In laboratory studies, it is possible to measure potential and current fluctuations simultaneously. In this approach, two electrodes of the same material are coupled through a zero-resistance ammeter (ZRA). Current fluctuations are measured using the ZRA, while potential fluctuations are measured with a high-impedance voltmeter between the coupled electrodes and a reference electrode that could be a stable reference electrode, such as an SCE or a third electrode of the same material as the two test electrodes. Simultaneous collection of potential and current EN data allows for analysis in time and frequency domains, and analysis of EN data in the time domain results in values of the mean potential (E_{coup}) and mean current (I_{coup}) of the coupled electrodes. In addition, the standard deviations σV and σI of the potential and current fluctuations, respectively, and noise resistance $R_n = \sigma V / \sigma I$ are obtained. The spectral noise resistance R_{sn}^o is defined as the dc limit of the spectral noise plot. Localization index (LI) $= \sigma I / rmsI$, where $rmsI$ is the root-mean-square value of the current fluctuations, can also be determined. Bertocci et al. (1997a, b) have described methods for data analysis. The main application of EN data has been in corrosion monitoring (Eden, 1998) and will be discussed in Chapter 5.

Microsensors

Lee and deBeer (1995) described ideal microsensors as having the following qualities: small tip diameters to prevent distortion of the local environment, small sensor surfaces for optimal spatial resolution, low noise levels, stable signal, high selectivity, and strength to resist breakage. Specific microelectrode design has been described elsewhere (Lewandowski et al., 1992).

Lewandowski et al. (1988) and Lewandowski (2000) used microelectrodes to determine the oxygen concentration around a microcolony (Figure 4-2). The microcolony was anoxic in the middle but oxygen was detected at the bottom, demonstrating transport via channels and voids in addition to diffusion. Microsensors have been used to develop profiles of mixed species biofilms. Figure 2-6 (Chapter 2)

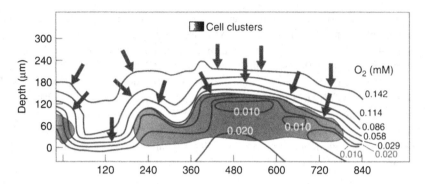

FIGURE 4-2   Oxygen concentration around a microcolony measured by microprobe. (Lewandowski, 2000. © NACE International.)

shows the concentration profiles of sulfide, oxygen, and pH in a biofilm accumulated on the surface of a carbon steel coupon (Lee et al., 1993). Concentration of sulfide is highest near the metal surface where iron sulfide forms quickly and covers the steel surface if both ferrous and sulfide ions are available.

Dickinson et al. (1996b) also used microelectrodes to measure dissolved oxygen (DO), H_2O_2, and local E_{corr} within biofouling deposits on S31603 stainless-steel surfaces exposed to river water to further resolve the interfacial chemistry that resulted in ennoblement. Their study indicated that elevated levels of dissolved oxidants did not cause ennoblement. Measurements at various heights above the substratum and at numerous sites over the coupon surface showed H_2O_2 concentrations <2 µM and no significant variation in E_{corr} for the stainless-steel microelectrode at any site. DO profiles in the same regions showed saturation levels at all sites. Representative profiles are shown in Figure 4-3.

Scanning Vibrating Electrode Techniques

Scanning vibrating electrode techniques (SVET) provide a sensitive means of locating local anodic and cathodic currents (vibrating microreference electrode) and potential distributions (Kelvin probe) associated with corrosion. Scanning vibrating electrode techniques are nondestructive to biofilms and their components and can provide qualitative and quantitative data. Franklin et al. (1990, 1991b, 2000) used autoradiography of bacteria following uptake of ^{14}C-acetate to locate bacteria, and SVET to locate anodic and cathodic currents on colonized steel (UNS G10200) surfaces (Figure 4-4a–c). They demonstrated that pit propagation in carbon steel exposed to a phosphate-containing electrolyte required either stagnant conditions or microbial colonization of anodic regions. In sterile, continuously aerated medium, pits initiated and repassivated, while in the absence of aeration, pits initiated and propagated. Pit propagation was observed in continuously aerated medium inoculated with a heterotrophic bacterium. Sites of anodic activity coincided with sites of bacterial activity. Results suggest that bacteria may preferentially attach to the

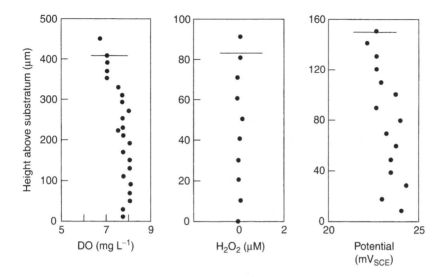

FIGURE 4-3   Microelectrode profiles in biofouling deposits on ennobled 316L stainless-steel (UNS S31603) coupons. Horizonal line indicates biofouling/solution interface. (Dickinson et al., 1996. © NACE International.)

corrosion products formed over corrosion pits. Biofilms over anodic sites may create stagnant conditions within pits, resulting in pit propagation. Scanning vibrating electrode techniques are successfully used to monitor early phases of MIC but are usually complemented by other investigative methods for the overview of corrosion processes.

Capacitance

Dickinson et al. (1996b) monitored capacitance (C) and E_{corr} values for S31603 stainless-steel exposed to river water over 50 days as a broad measure of near-surface changes that may reflect changes in surface oxides. C was determined using the galvanostatic transient method. A 25 s, constant-current pulse (2–10 μA cm^{-2}) was applied to stainless-steel coupons to produce an overvoltage–time response. Applied current (I_{app}) was chosen to generate less than -10 mV cathodic overvoltage (η). Typical η was in the range of -2 to -8 mV. After amplification through a high-impedance differential amplifier, signals were recorded at 10 ms intervals by computer. C was determined by nonlinear fitting as:

$$\eta = I_{app}R_p\,(1-\exp[t/R_pC]) \tag{4-1}$$

where t is the time, assuming a simple parallel combination of R_p and C as the model for the electrode–solution interface. In most cases, this produced an acceptable fit, although introduction of a second time constant was required for some measurements

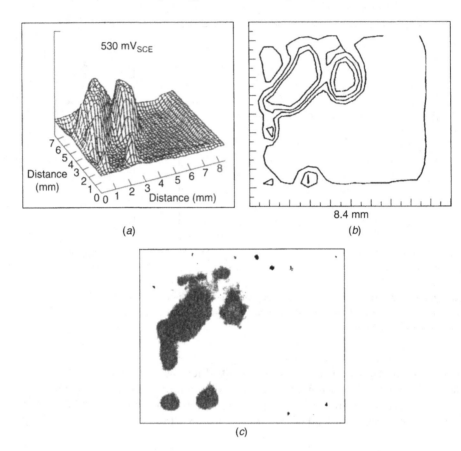

(a)

8.4 mm

(b)

(c)

FIGURE 4-4 a–c   Bacteria labeled by incubating with ^{14}C-acetate prior to exposure to corroding sample. Bacteria were associated primarily with anodic sites: (a) potential field map, (b) contour map of potential fields, and (c) autoradiogram. (Reprinted from Franklin et al., 2000, with permission from Taylor and Francis Ltd. http://www.tandf.uk/journals.)

to fit the data accurately. Close proximity of the working and reference electrodes and the small I_{app} made corrections for uncompensated resistance unnecessary.

Figure 4-5 shows the evolution of cathodic response to a 10 nA cm^{-2}, 25 s galvanostatic pulse for a representative sample as E_{corr} increased during the river water exposures. The overvoltage–time response fit to Eq. (4-1) by nonlinear regression was used to determine R_p and C. The 10 ms sampling interval during measurements was reflected by the continuous data curves, while filled circles indicate individual points generated from the computed R_p and C values. Data were fit by the simple parallel resistance–capacitance (RC) model used to generate Eq. (4-5). Initial values of R_p and C were in the ranges 0.5 to 1.5 M Ω cm^2 and 35 to 55 μF cm^{-2}, respectively, in good agreement with results obtained for stainless steel by Mansfeld et al. (1992) using electrochemical impedance spectroscopy

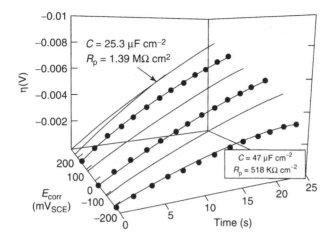

FIGURE 4-5 Evolution of cathodic response to 10 μA cm^{-2} galvanostatic pulse as E_{corr} increases during biofouling of 316L stainless steel (UNS S31603) in freshwater. Data shown by solid lines and circles indicate points generated from curve fit to Eq. (4-1). (Dickinson et al., 1996b. © NACE International.)

techniques. The steady decrease in final η with increasing E_{corr} shown in Figure 4-5 likely reflected development of increasing cathodic Tafel slope during ennoblement and appeared as an increase in the value of R_p. A corresponding decrease in C for data from Figure 4-5 is shown in Figure 4-6a. In contrast, Figure 4-6b shows constant C–t behavior for a coupon that exhibited little change in E_{corr} during exposure.

To determine if the difference in C behavior for coupons in Figures 4-6a, b was due to a difference in biofouling accumulation, the extent of coverage and morphology of biofouling on the coupons was assessed by microscopic examination at the end of the exposure period. Biofouling on the two coupons was indistinguishable. In summary, neither the duration of exposure nor the mere presence of biofouling could account for the decrease in C shown in Figure 4-6a.

C values for a nonexposed coupon prepared in a manner identical to the exposed coupons were measured, while the coupon was polarized by constant anodic current to determine if changes in C resulted solely from increased potential. For each potential, polarizing current was applied for approximately 1 h until the potential drift was <2 mV h^{-1} before measuring for C values. Data indicated by open squares in Figure 4-7 show C values were independent of potential over the range -100 to 200 mV$_{SCE}$. This finding was in agreement with results obtained by Popat and Hackerman (1961) for 302 stainless-steel (UNS S30200) in 0.1 M Na$_2$SO$_4$ at neutral pH.

Dickinson et al. (1996b) determined a strong correlation between C and ennobled E_{corr} (Figure 4-7) of S31603 coupons. C is expressed as a fraction of initial C (C_{init}) to account for variation in true surface area of the coupons. Origin of the observed correlation was not established, and the relation was not interpreted as causal.

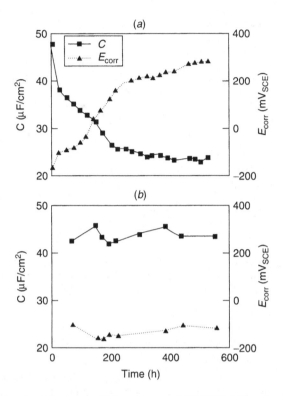

FIGURE 4-6 *a*, *b* C and E_{corr} for two 316L stainless-steel (UNS S31603) coupons during biofouling in fresh river water: (*a*) E_{corr} ennobled during exposure, and (*b*) E_{corr} nearly constant during exposure. (Dickinson et al., 1996b. © NACE International.)

Dual-Cell Technique

The dual cell, split cell, or biological battery allows continuous monitoring of changes in corrosion rates due to the presence of a biofilm. Two identical electrochemical cells are separated biologically by a semipermeable membrane. The two working electrodes are connected to a ZRA or a potentiostat set to an applied potential of 0 mV. Bacteria are added to one of the two cells, and the sign and magnitude of the resulting galvanic current are monitored to determine details of the corrosive action of the bacteria. Little et al. (1997) used a dual cell to demonstrate MIC of carbon steel (UNS G10100) in the presence of an iron-reducing bacterium. The dual-cell technique does not provide a means to calculate corrosion rates, but helps monitor changes due to the presence of a biofilm.

Dexter and LaFontaine (1998) evaluated galvanic corrosion of copper (UNS C11000), carbon steel (UNS G1018), 3003 aluminum (UNS A93003), and zinc (UNS Z32121) anodes coupled to cathode panels of passive alloy AL6XN (UNS N08367). Natural marine microbial biofilms were allowed to form on the cathode surface. On the control, formation of the biofilm was prevented. Corrosion of

C11000, G1018, and A93003 anodes was significantly higher when connected to cathodes on which biofilms were allowed to grow naturally. The largest effect was measured for C11000 anodes. As shown in Figure 4-8, the galvanic current for C11000 corrosion as measured with a ZRA was 2–3 decades higher for C11000 couples with biofilms on the cathode than for the corresponding control couples. Weight

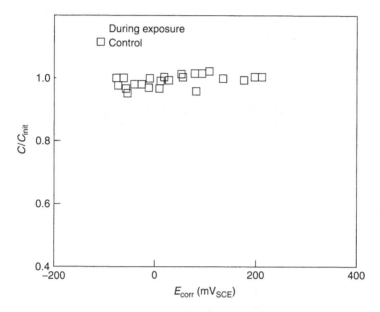

FIGURE 4-7 C/C_{init} ratio versus E_{corr} for six 316L stainless-steel (UNS S31603) coupons during biofouling in freshwater and for control coupon. (Dickinson et al.,1996b. © NACE International.)

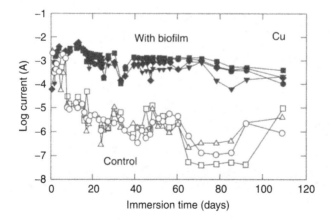

FIGURE 4-8 Corrosion currents for galvanic couples of UNS C11000 copper anodes versus UNS NO8367 stainless-steel cathodes with and without (control) the influence of natural marine microbial biofilms. (Dexter and LaFontaine, 1998. © NACE International.)

loss for C11000 anodes was also significantly higher when coupled to cathodes with biofilms than on control coupons.

Similar results were found for G1018 and A93003 anodes. Average corrosion currents and weight losses were 5–8 times higher when biofilms were allowed to grow on the cathode surface than for the control. Galvanic currents measured for A93003 are shown in Figure 4-9, and the A93003 anodes are shown in Figure 4-10a–c.

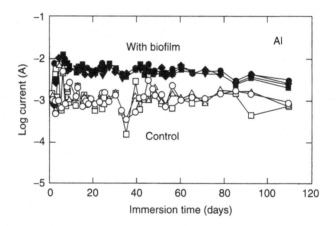

FIGURE 4-9   Corrosion currents for galvanic couples of UNS A93003 aluminum-alloy anodes versus UNS NO8367 stainless-steel cathodes with and without (control) the influence of natural marine microbial biofilms. (Dexter and LaFontaine, 1998. © NACE International.)

FIGURE 4-10a–c   UNS A93003 aluminum anodes after galvanic corrosion test and removal of corrosion products: (a) coupled to UNS NO8367 stainless-steel cathode with natural microbial biofilm, (b) coupled to control UNS NO8367 stainless-steel cathode without biofilm, and (c) as originally prepared prior for immersion. (Dexter and LaFontaine, 1998. © NACE International.)

TECHNIQUES REQUIRING A SMALL EXTERNAL SIGNAL

Polarization Resistance Technique

Polarization resistance (R_p) techniques can be used to continuously monitor the instantaneous corrosion rate of a metal. Mansfeld (1976) provided a thorough review of the use of the polarization resistance technique for the measurement of corrosion currents. R_p is defined as

$$R_p = (dE/di)i = 0 \qquad (4\text{-}2)$$

R_p is the slope of a potential (E) versus current density (i) curve at E_{corr}, where $i = 0$. Corrosion current density (i_{corr}) is calculated from R_p by

$$i_{corr} = B/R_p, \qquad (4\text{-}3)$$

where

$$B = b_a b_c / 2.303 (b_a + b_c). \qquad (4\text{-}4)$$

The exact calculation of i_{corr} for a given time requires simultaneous measurements of R_p and anodic and cathodic Tafel slopes (b_a and b_c) (Shih and Mansfeld, 1992). Computer programs have been developed for determining the precise values of i_{corr} according to Eqs. (4-3) and (4-4). Experimental values of R_p (R'_p) contain a contribution from the uncompensated solution resistance (R_u):

$$R'_p = R_p + R_u \qquad (4\text{-}5)$$

Applications of R_p techniques have been reported by King et al. (1986) in a study of the corrosion behavior of grey and ductile iron pipes (undesignated) in environments containing SRB. In a similar study, Kasahara and Kajiyama (1986) used R_p measurements with compensation of the ohmic drop and reported results for active and inactive SRB. Nivens et al. (1991) calculated the corrosion current density from experimental R_p data and Tafel slopes for S30400 stainless steel exposed to a seawater medium containing the non-SRB *Vibrio natriegens.*

A simplification of the polarization resistance technique is the linear polarization technique in which it is assumed that the relationship between E and i is linear in a narrow range around E_{corr}. Usually, only two points (E, i) are measured, and B is assumed to have a constant value of about 20 mV. This approach is used in field tests and forms the basis of commercial corrosion rate monitors. R_p can also be determined as the dc limit of electrochemical impedance. Mansfeld et al. (1991) used the linear polarization technique to determine R_p for unspecified carbon steel sensors embedded in concrete exposed to a sewer environment for about 9 months. One sensor was periodically flushed with sewage in an attempt to remove the sulfuric acid produced by sulfur-oxidizing bacteria within a biofilm, while another sensor was used as a control. A data logging system collected R_p at

10 min intervals simultaneously for the two corrosion sensors and two pH electrodes placed at the concrete surface. Figure 4-11*a, b* shows the cumulative corrosion loss (ΣINT) obtained by integration of the $1/R_p$ time curves as

$$\Sigma INT = \int \frac{dt}{R_p} \qquad (4\text{-}6)$$

A qualitative measure of the corrosion rate can be obtained from the slope of the curves in Figure 4-11. ΣINT is given in units of s Ω^{-1}. Owing to the presence of the uncompensated ohmic resistance and lack of values for Tafel slopes, data in Figure 4-11 should be viewed as indicative of significant changes in corrosion rates. Corrosion loss remained low during the first 2 months, followed by a large increase for both flushed samples and controls. Increased corrosion rate occurred when the surface pH reached values of 1 or less. Total corrosion loss as determined from integrated R_p data was less for the control than for the flushed sample.

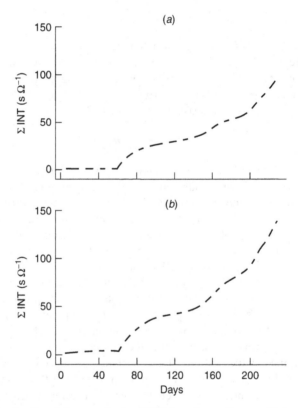

FIGURE 4-11*a, b* Cumulative corrosion loss ΣINT for carbon steel (unspecified) sensors embedded in concrete and exposed to a sewer bypass: (*a*) control sample, and (*b*) flushed sample. (Mansfeld et al., 1991. © NACE International.)

Lee et al. (2004) used linear polarization measurements to demonstrate that corrosion of carbon steel (G10200) was more aggressive in stagnant anaerobic seawater than in stagnant aerobic seawater over a 396-day exposure period. Coupons were labeled as to placement within exposure chambers with row 1 vertically oriented near the top, rows 2 and 3 vertically oriented at intermediate levels, and row 4 horizontally oriented at the bottom. R_p was calculated for individual coupons and averaged by row and exposure condition; $1/R_p$ was plotted against exposure time (Figure 4-12).

Significant errors in the calculation of corrosion rates can occur for electrolytes of low conductivity or systems with very high corrosion rates (low R_p) if a correction for R_u is not applied. Corrosion rates will be underestimated in these cases. Additional problems can arise from the effects of the sweep rate used to determine R_p according to Eq. (4-2). If the sweep rate is too high, the experimental value of R_p will be too low and the calculated corrosion rate will be too high. For localized corrosion, experimental R_p data should be used as a qualitative indication that rapid corrosion is occurring. Large fluctuations of R_p with time are often observed for systems undergoing pitting or crevice corrosion. R_p data are meaningful for general or uniform corrosion but less so for localized corrosion, including MIC. Additionally, the use of Stern–Geary theory (where corrosion rate is inversely proportional to R_p

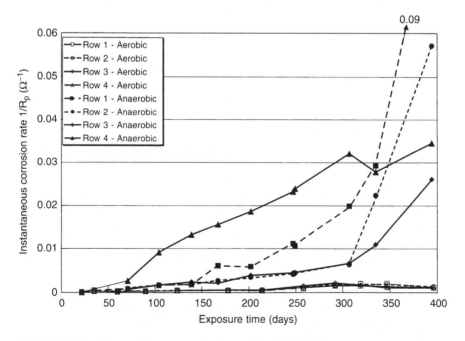

FIGURE 4-12 $1/R_p$ (corrosion rate) over time (days) for UNS G10200 samples exposed to stagnant aerobic and anaerobic Key West seawater over a 396-day period. Rows 1, 2, 3 — vertically orientated samples with Row 1 being at the top of the tank, Row 3 towards the bottom and Row 2 between the two. Row 4 — horizontally orientated sample at the bottom of the tank. Average values displayed. (Lee et al., 2004.)

at potentials close to E_{corr}) is valid for conditions controlled by electron transfer, but not for diffusion-controlled systems as frequently found in MIC.

Electrochemical Impedance Spectroscopy

Electrochemical impedance spectroscopy (EIS) techniques record impedance data as a function of the frequency of an applied signal at a fixed potential. A large frequency range (65 kHz–1 MHz) must be investigated to obtain a complete impedance spectrum. Dowling et al. (1988) and Franklin et al. (1991a) demonstrated that the small signals required for EIS do not adversely affect the numbers, viability, and activity of microorganisms within a biofilm. Electrochemical impedance spectroscopy data may be used to determine R_p, the inverse of corrosion rate. Electrochemical impedance spectroscopy is commonly used for steady-state conditions (uniform corrosion); however, sophisticated models have been developed for localized corrosion (Kendig et al., 1983; Mansfeld et al., 1982). Several reports have been published in which EIS has been used to study the role of SRB in corroding buried pipes (King et al., 1986; Kasahara and Kajiyama, 1986, 1991) and reinforced concrete (Iverson, 1968; Lee et al., 1993). Formation of biofilms and calcareous deposits on three stainless steels (UNS S30400, S31600, and N08367) and titanium (ASTM grade 2, UNS R50400) during exposure to natural seawater was monitored using EIS and surface analysis (Figure 4-13) (Mansfeld et al., 1990).

Dowling et al. (1989) studied the effects of MIC on S31603 stainless-steel weldments in artificial seawater using EIS and small-amplitude cyclic voltammetry. They concluded from the frequency dependence of the impedance data that two relaxations were associated with the as-welded inoculated surface, while only one time constant was detected in the Bode plots for the as-welded sterile polished surfaces, speculating that the occurrence of a second time constant was due to the development of pits. Dowling et al. (1988) used EIS to study the corrosion behavior of carbon steels (UNS G10200 and ASTM A284 [E24]) affected by bacteria and attempted to determine R_p from the EIS data. Ferrante and Feron (1991) used EIS data to conclude that the material composition of steels was more important for MIC resistance than bacterial population, incubation time, sulfide content, and other products of bacterial growth. Jones et al. (1991) used EIS to determine the effects of several mixed microbiological communities on the protective properties of epoxy top coatings over zinc-primed steel (UNS G41400). Spectra for the control remained capacitive, indicating intact coatings, while spectra for five of six samples exposed to mixed cultures of bacteria indicated corrosion and delamination (Figure 4-14). Figure 4-15 shows EIS Bode plots for carbon steel (UNS G10200) immersed in natural oilfield-produced water with (Figure 4-15b) and without (Figure 4-15a) added nutrients to support SRB growth (Luo et al., 1994). Few changes resulted in the supplemented waters while significant alterations were observed in those without supplements. Figure 4-15a shows a greater increase in the phase angle at lowest frequencies, suggesting formation of a thicker biofilm.

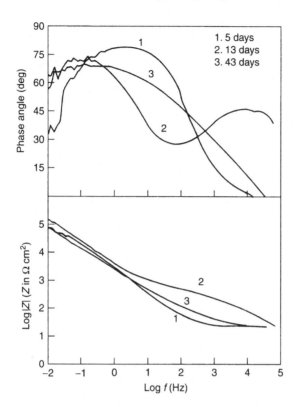

FIGURE 4-13   Impedance spectra for polarized 316 stainless steel (UNS S31600) after exposure periods of 5, 13, and 43 days. EIS data taken at E_{corr}. (Mansfeld et al., 1990.)

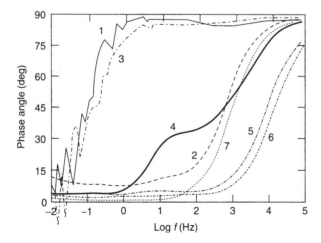

FIGURE 4-14   EIS Bode plot for epoxy coating over zinc-primed steel (UNS G41400) exposed for 2 months to uninoculated medium (curve 1) and six mixed cultures of facultative anaeroves (curves 2–7). (Jones et al., 1991.)

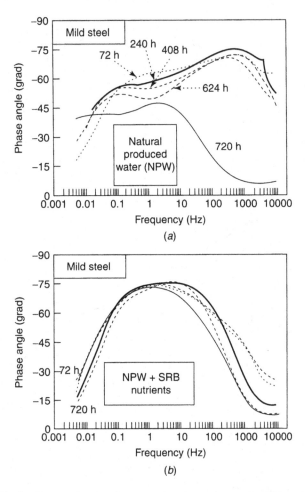

FIGURE 4-15a, b   Phase angle versus frequency plots from carbon steel at different immersion times in (a) natural oilfield-produced water, and (b) the same water with added SRB nutrients. (Luo et al., 1994. © NACE International.)

LARGE SIGNAL POLARIZATION

Large signal polarization techniques require potential scans ranging from several hundred millivolts to several volts. Large signal polarization is applied to obtain potentiostatic or potentiodynamic polarization curves as well as pitting scans. Polarization curves can be used to determine i_{corr} by Tafel extrapolation, while mass transport-related phenomena can be evaluated based on the limiting current density [(c.d.) i_{lim}]. Mechanistic information can be obtained from experimental values of b_a and b_c. For metal/electrolyte systems for which an active–passive transition occurs, the passive properties can be evaluated based on the critical potential (E_{crit}) and critical c.d. i_{crit} for passivation and the passive c.d. i_{pass}. Pitting scans are used to determine E_{pit} and the protection potential (E_{prot}).

A disadvantage of large signal polarizations is their destructive nature, that is, the irreversible changes of surface properties due to application of large anodic or cathodic potentials. Choice of scan rate is important in MIC studies to reduce effects on biofilm structure and character. The faster the scan rate, the less the impact on microbial activities. Recording polarization curves provides an overview of reactions for a given corrosion system — charge transfer or diffusion controlled reactions, passivity, transpassivity, and localized corrosion phenomena.

Numerous investigators have used polarization curves to determine the effects of microorganisms on the electrochemical properties of metal surfaces and the resulting corrosion behavior. In most of these studies, comparisons have been made between polarization curves in sterile media and those obtained in the presence of bacteria and fungi. Figure 4-16 shows current density versus potential curves for corrosion systems with and without bacteria (Schmitt, 1997). The complexity of the BS-251 (non-UNS designation) naval brass/polluted seawater interface was demonstrated by Deshmukh et al. (1988), who discussed the influence of sulfide pollutants based on results obtained with potentiodynamic polarization curves.

Keresztes et al. (1998) used measurements of E_{corr}, R_p, and potentiostatic polarization measurements to obtain corrosion rates (see Figure 4-17). Culture media containing sulfide of both biogenic and chemical origin were used to determine the effects of metal-sulfide layers. Biocides were used to inhibit bacterial metabolic activity. An atomic force microscope was used to image the topography of sulfide layers. They concluded that SRB produced continuous and localized sulfide,

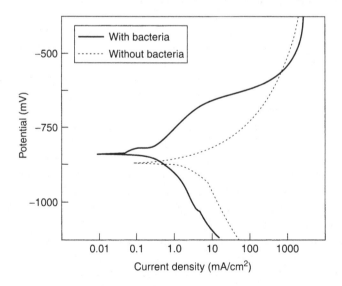

FIGURE 4-16   Current density versus potential curves of a corrosion system, with and without bacteria, obtained by large signal polarization. (Reprinted from Schmitt, 1997, with permission of Wiley–VCH, Weinheim, Germany, and the author.)

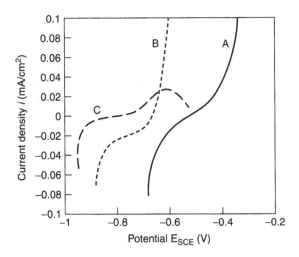

FIGURE 4-17   Polarization curves measured on carbon steel (containing 99.57% Fe, 0.05% Si, 0.05% C, 0.3% Mn, 0.01% S, 0.02% P) in the solution of biogenic and chemical sulfide. A, blank; B, biogenic sulfide; and C, chemical sulfide. (Reprinted Kereszetes et al., 1998 with permission from Elsevier.)

regenerating anodic sites and — in the case of iron — activating cathodic sites in the vicinity of the anodes.

CONCENTRIC RING ELECTRODES

Concentric ring electrodes have been used to examine the impact of physical separation of anodes and cathodes on MIC. Polarization was employed to precondition the electrodes for preferential regions of microbiological activity. However, motivation for polarization differed by two distinct categories: (1) mimic corrosion sustained by biofilm formation (Angell et al., 1995; Campaignolle and Crolet, 1997), and (2) monitor biofilm formation through measurement of electrochemical parameters (Licina and Nekoksa, 1995). Monitoring devices based on physical separation of anode and cathode will be discussed in Chapter 5.

Champaignolle and Crolet (1997) used a concentric ring carbon steel (G10200) electrode to examine stabilization of pitting corrosion by SRB (Figure 4-18). The anode was concentric to, and separated from, the cathode (4.87 cm^2) by a polytetrafluoroethylene (PTFE) spacer. Pitting was induced by passage of an 11 μA cm^{-2} current density to a small (0.031 cm^2) anode. Current was applied for 72 h either during or after microbial colonization. Once the applied current was removed, the resultant galvanic current flowing between the anode and cathode was monitored by a ZRA. They found that a current was maintained in the presence of a microbial consortium. A very small current was measured in a sterile control. The authors state that the concentric ring electrode provides a technique by which MIC can be studied and is not intended to represent any natural situation.

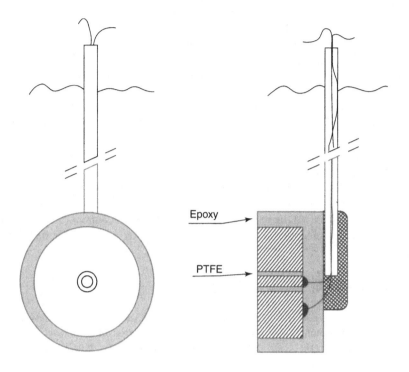

FIGURE 4-18 Schematic of concentric ring electrodes. (Campaignolle and Crolet, 1997. © NACE International.)

Angell et al. (1995) used a concentric ring S30400 stainless-steel electrode to demonstrate that a consortium of SRB and a *Vibrio* sp. maintained a galvanic current between the anode and cathode. A current density of 1.5 mA cm^{-2} was applied to the anode versus the cathode for 48 h to induce corrosion in deoxygenated nutrient-enriched synthetic seawater. During this preconditioning period, the test medium was inoculated with *Desulfovibrio vulgaris* while a sterile condition was monitored as the control. In the presence of *D. vulgaris*, a residual galvanic current was obtained while no current was measured in the sterile control. SRB were shown to stabilize pitting corrosion of S30400 stainless steel by maintaining a stable galvanic current between a local anode and a cathodic surrounding. In the absence of SRB, the current was negligible. The authors noted that sustained galvanic corrosion required both anode and cathode colonization by SRB. However, corrosion rates were still far from the observed rates in some field results.

SUMMARY

Electrochemical techniques have been used to demonstrate MIC for a large number of metals and alloys in a large number of environments. Often, corrosion rates as determined from polarization resistance employing DC and/or AC techniques have been compared for abiotic and biotic environments. While this approach has increased the general knowledge concerning the occurrence of MIC under a variety of circumstances, very little mechanistic information has been obtained.

Most corrosion reactions in MIC are of a very localized nature, making use of electrochemical techniques, which in general produce an average signal over the entire surface, difficult. In this respect, the use of electrochemical techniques in the evaluation of MIC is limited to the same degree as all studies of localized corrosion phenomena. The use of microsensors for the determination of local chemistries under a biofilm and the potential or current scanning techniques discussed in this chapter should provide additional data for mechanistic interpretation.

In an area of corrosion as complicated as MIC, a single experimental technique cannot provide all the answers needed for the establishment of mechanisms. As in other fields of corrosion, multiple techniques are required for mechanistic studies.

REFERENCES

Angell P, Luo J-S, White D-C (1995). Studies of the reproducible pitting of 304 stainless steel by a consortium containing sulphate-reducing bacteria. *International Conference on Microbially Influenced Corrosion*. New Orleans, LA: NACE International, pp. 1/1–1/10.

Bertocci U, Gabrielli C, Huet F, Keddam M (1997a). Noise resistance applied to corrosion measurements. I. Theoretical analysis. *J. Electrochem. Soc.*, **144**: 31.

Bertocci U, Gabrielli C, Huet F, Keddam M, Rousseau P (1997b). Noise resistance applied to corrosion measurements. II. Experimental tests. *J. Electrochem. Soc.*, **144**: 37.

Campaignolle X, Crolet JL (1997). Method for studying stabilization of localized corrosion on carbon steel by sulfate-reducing bacteria. *Corrosion*, **53**: 440–447.

Chandrasekaran P, Dexter SC (1993). Mechanism of potential ennoblement on passive metals by seawater biofilms. *CORROSION/93*, Paper No. 493. Houston TX: NACE International.

Deshmukh MB, Akhtar I, De CP (1988). Influence of sulphide pollutants of bacterial origin on corrosion behaviour of naval brass. In: *2nd International Symposium on Industrial and Oriented Basic Electrochemistry*. SAEST, India. Oxford: IBH Publishing.

Dexter SC, LaFontaine JP (1998). Effect of natural marine biofilms on galvanic corrosion. *Corrosion*, **54**(11): 851–861.

Dickinson WH, Caccavo F, Lewandowski Z (1996a). The ennoblement of stainless steel by manganic oxide biofouling. *Corros. Sci.*, **38**(8): 1407.

Dickinson WH, Lewandowski Z (1996). Manganese biofouling and the corrosion behavior of stainless steel. *Biofouling*, **10**: 79–93.

Dickinson WH, Lewandowski Z, Geer RD (1996b). Evidence for surface changes during ennoblement of type 316L stainless steel: dissolved oxidant and capacitance measurements. *Corrosion*, **52**(12): 910.

Dowling NJE, Franklin M, White DC, Lee CH, Lundin C (1989). The effect of microbiologically influenced corrosion on stainless steel weldments in artificial seawater. *CORROSION/89*, Paper No. 187. Houston, TX: NACE International.

Dowling NJE, Guezennec J, Lemoine ML, Tunlid A, White DC (1988). Analysis of carbon steels affected by bacteria using electrochemical impedance and direct current techniques. *Corrosion*, **44**(12): 869.

Eashwar M, Maruthamuthu S, Palanichamy S, Balakrishnan K (1995a). Sunlight irradiation of seawater eliminates ennoblement causation by biofilms. *Biofouling*, **8**: 215–221.

Eashwar M, Maruthamuthu S, Sathiyanarayanan S, Balakrishnan K (1995b). The ennoblement of stainless alloys by marine biofilms: the neutral pH and passivity enhancement model. *Corros. Sci.*, **37**(8): 1169–1176.

Eashwar M, Subramanian G, Chandrasekaran P, Balakrishnan K (1992). Mechanism for barnacle-induced crevice corrosion in stainless steel. *Corrosion*, **48**(7): 608–612.

Eden DA (1998). Electrochemical noise — The first two octaves. *CORROSION/98*, Paper No. 386. Houston, TX: NACE International.

Ferrante V, Feron D (1991). Microbially influenced corrosion of steels containing molybdenum and chromium: a biological and electrochemical study. In: Dowling NJ, Mittleman MW, Danko JC, (eds) *Microbially Influenced Corrosion and Biodeterioration*. Knoxville, TN: University of Tennessee, pp. 3-55–3-63.

Franklin MJ, Nivens DE, Guckert JB, White DC (1991a). Effect of electrochemical impedance spectroscopy on microbial biofilm cell numbers, viability, and activity. *Corrosion*, **47**(7): 519.

Franklin MJ, White DC, Isaacs HS (1990). The use of current density mapping in the study of microbial influenced corrosion. *CORROSION/90*, Paper No. 104. Houston, TX: NACE International.

Franklin MJ, White DC, Isaacs HS (1991b). Pitting corrosion by bacteria on carbon steel, determined by the vibrating electrode technique. *Corros. Sci.*, **32**(9): 945.

Franklin MJ, White DC, Little B, Ray R, Pope R (2000). The role of bacteria in pit propagation of carbon steel. *Biofouling*, **15**(1-3): 13–23.

Iverson WP (1968). Transient voltage changes produced in corroding metals and alloys. *J. Electrochem. Soc.*, **115**: 617.

Johnsen R, Bardal E (1985). Cathodic properties of different stainless steels in natural seawater. *Corrosion*, **41**(5): 296.

Jones J, Walch M, Mansfeld F (1991). Microbial electrochemical studies of coated steel exposed to mixed microbial communities. *CORROSION/91*, Paper No. 108. Houston, TX: NACE International.

Kasahara K, Kajiyama F (1986). Role of sulfate-reducing bacteria in the localized corrosion of buried pipes. In: Dexter SC (ed) *Biologically Induced Corrosion*. Houston, TX: NACE International, pp. 172–183.

Kasahara K, Kajiyama F (1991). Electrochemical aspects of microbiologically influenced corrosion on buried pipes. In: Dowling NJ, Mittelman MW, Danko JC (eds) *Microbially Influenced Corrosion and Biodeterioration*. Knoxville, TN: University of Tennessee, pp. 2-33–2-39.

Kendig M, Mansfeld F, Tsai S (1983). Determination of the long-term corrosion behavior of coated steel with A.C. impedance measurements. *Corros. Sci.*, **23**(4): 317.

Kereszetes ZS, Telegdi J, Beczner J, Kalman E (1998). The influence of biocides on the microbiologically influenced corrosion of mild steel and brass. *Electrochim. Acta*, **43**(1-2): 77–85.

King RA, Skerry BS, Moore DCA, Stott JFD, Dawson JL (1986). Corrosion behaviour of ductile and grey iron pipes in environments containing sulphate-reducing bacteria. In: Dexter SC (ed) *Biologically Induced Corrosion*. Houston, TX: NACE International, pp. 83–91.

Lee JS, Ray RI, Lemieux EJ, Falster AU, Little BJ (2004). An evaluation of carbon steel corrosion under stagnant seawater conditions. *Biofouling*, **20**: 237–247.

Lee W, deBeer D (1995). Oxygen and pH microprofiles above corroding mild steel covered with a biofilm. *Biofouling*, **8**: 273.

Lee W, Lewandowski Z, Morrison M, Characklis WG, Avsi R, Nielsen P (1993). Corrosion of mild steel underneath aerobic biofilms containing sulfate-reducing bacteria. Part II: At high dissolved oxygen concentration. *Biofouling*, **7**: 217.

Lewandowski Z (2000). MIC and biofilm heterogeneity. *CORROSION/2000*, Paper No. 400. Houston, TX: NACE International.

Lewandowski Z, Characklis W, Lee W, Little B (1992). Construction and application of iridium oxide pH microelectrode. *J. Bioeng. Biotech.*, **40**: 601–608.

Lewandowski Z, Lee WC, Characklis WG, Little B (1988). Dissolved oxygen and pH microelectrode measurements at water-immersed metal surfaces. *CORROSION/88*, Paper No. 93. Houston, TX: NACE International.

Licina GJ, Nekoksa G (1995). On-line monitoring of biofilm formation for the control and prevention of microbially influenced corrosion. *International Conference on Microbially Influenced Corrosion*. New Orleans, LA: NACE International, pp. 42/41–42/10.

Linhardt P (1994). Manganese oxidizing bacteria and pitting of turbine components made of CrNi steel in a hydroelectric power plant. *Werkst. Korros.*, **45**(2): 79–83.

Linhardt P (1996). Pitting of stainless steel in freshwater influenced by manganese oxidizing microorganisms. *DECHEMA Monographs*, p. 133.

Little B J, Wagner P A, Hart K R, Ray R I, Lavoie D M, Nealson K, Aguilar C (1997). The role of metal-reducing bacteria in microbiologically influenced corrosion. *CORROSION/97*, Paper No. 215, Houston, TX: NACE International.

Little BJ, Wagner PA (2001). Application of electrochemical techniques to the study of MIC. In: *Modern Aspects of Electrochemistry*. New York: Kluwer Academic/Plenum Publishers.

Luo JS, Angell P, White DC, Vance I (1994). MIC of mild steel in oilfield-produced water. *CORROSION/94*, Paper No. 265. Houston, TX: NACE International.

Mansfeld F (1976). The polarization resistance technique for measuring corrosion currents. In: Fontana MG, Staehle RW (eds) *Advances in Corrosion Science and Technology*, Vol. 6. New York: Plenum Press, pp. 163–262.

Mansfeld F, Kendig M, Tsai S (1982). Evaluation of corrosion behavior of coated metals with AC impedance measurements. *Corrosion*, **38**: 478.

Mansfeld F, Shih H, Postyn A, Devinny J, Islander R, Chen CL (1991). Corrosion monitoring and control in sewer pipes. *Corrosion*, Paper No. 47. Houston, TX: NACE International.

Mansfeld F, Tsai R, Shih H, Little B, Ray R, Wagner P (1990). Results of exposure of stainless steels and titanium to natural seawater. *CORROSION/90*, Paper No. 109; Houston, TX: NACE International.

Mansfeld F, Tsai R, Shih H, Little B, Ray R, Wagner P (1992). An electrochemical and surface analytical study of stainless steels and titanium exposed to natural seawater. *Corros. Sci.*, **33**(3): 445.

Mollica A, Trevis A (1976). The influence of the microbiological film on stainless steels in natural seawater. In: Pins J-L (ed) *Proceedings of the 4th International Congress on Marine Corrosion and Fouling*. Antibes, France, p. 351.

Nivens DE, Chambers JQ, White DC (1991). Non-destructive monitoring of microbial biofilms at solid–liquid interface using on-line devices. In: Dowling NJ, Mittelman MW, Danko JC (eds) *Microbially Influenced Corrosion and Biodeterioration*. Knoxville, TN: University of Tennessee, pp. 5-47-5-56.

Olesen BH, Avci R, Lewandowski Z (2000a). Manganese dioxide as a potential cathodic reactant in corrosion of stainless steels. *Corros. Sci.*, **42**: 211–227.

Olesen BH, Nielsen PH, Lewandowski Z (2000b). Effect of biomineralized manganese on the corrosion behavior of C1008 mild steel. *Corrosion*, **56**: 80–89.

Popat PV, Hackerman N (1961). Electrical double layer capacity of passive iron and stainless steel electrodes. *J. Phys. Chem.*, **65**: 1201.

Renner M (1996). Scientific, engineering, and economic aspects of MIC on stainless steels application in the chemical process industry. *DECHEMA Monographs*, Vol. 133, pp. 59–70.

Ruppel DT, Dexter SC, Luther GW (2001). Role of manganese dioxide in corrosion in the presence of natural biofilm. *Corrosion*, **57**: 863–873.

Schmitt G (1997). Sophisticated electrochemical methods for MIC investigation and monitoring. *Mater. Corros.* **48**: 586.

Scotto V, Di Cinto R, Marcenaro G (1985). The influence of marine aerobic microbial film on stainless steel corrosion behaviour. *Corros. Sci.*, **25**: 185–194.

Shih H, Mansfeld F (1992). Software for quantitative analysis of polarization curves. In: Munn RS (ed) *ASTM STP 1154 Computer Modeling and Corrosion*. Philadelphia, PA: American Society for Testing and Materials, pp. 174–185.

Zhang H-J, Dexter SC (1995). Effect of biofilms on crevice corrosion of stainless steels in coastal seawater. *Corrosion*, **51**(1): 56–66.

Zhang X, Buchanan RA, Stansbury EE, Dowling NJE (1989). Electrochemical responses of structural materials to microbially influenced corrosion. *CORROSION/89*, Paper No. 512. Houston, TX. NACE International.

Chapter 5

Approaches for Monitoring Microbiologically Influenced Corrosion

INTRODUCTION

Many techniques claim to monitor microbiologically influenced corrosion (MIC). However, none has been accepted as an industry standard or as a recommended practice by the American Society for Testing and Materials or NACE International. Some techniques can detect a specific modification in the system due to the presence and activities of microorganisms (e.g., heat-transfer resistance, fluid friction resistance, and galvanic current) and assume something about the corrosion. Others measure some electrochemical parameter (e.g., polarization resistance and electrochemical noise) and assume something about the microbiology. With experience and knowledge of a particular operating system either can be an effective monitoring tool, especially for evaluating a treatment regime (biocides or corrosion inhibitors). Several investigators have attempted to simultaneously quantify corrosion and biofilm formation. The electrochemical techniques used in the monitoring systems described below have been described in the previous chapter.

The terms used to describe monitoring tools are real-time, in-line and side-stream, and on-line. Real-time refers to measurements that are available at the actual time of collection and are usually continuous, or nearly continuous. In-line and side-stream define the position of the monitor in the system where side-stream devices are installed in parallel to the main system, taking a portion of the flow under identical operating conditions. On-line indicates that monitoring occurs during normal system operation.

Microbiologically Influenced Corrosion By Brenda J. Little and Jason S. Lee
Published 2007 by John Wiley & Sons, Inc.

COUPON HOLDERS

Biofilm formation and corrosion can be monitored using metal coupons implanted either in-line or in a side-stream. Several types of coupon holders have been developed (Figure 5-1a–c). Typically, the coupons are removed periodically and used for weight loss, microbiological analyses, or microscopic examination.

A variety of biofilm monitoring devices have been developed based on the influence of biofilms on heat transport or fluid-friction resistance (Knudsen, 1981; Characklis, 1981; Zelver et al., 1985; Block and DiFranco, 1995). These devices are based on the following principles. Flow of water through a conduit is associated with a pressure drop. As biofilm accumulates on the wall of the conduit, the pressure drop increases. Biofilm development may result in unusually high fluid-friction resistance losses because of increased roughness. Conductive heat-transfer resistance results from the insulating layer formed by the biofilm and generally increases as the

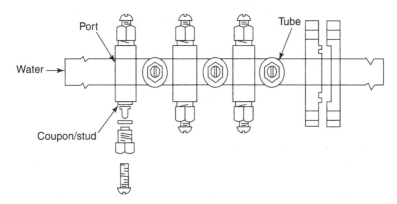

FIGURE 5-1a   Schematic of a Robbins device. (Provided by Tyler Research Corporation, Edmonton, Alberta, Canada.)

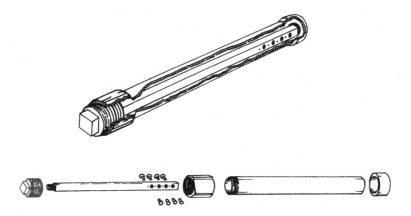

FIGURE 5-1b   RENAprobe ™ sampler device and carrier. (Videla et al., 1990. © NACE International.)

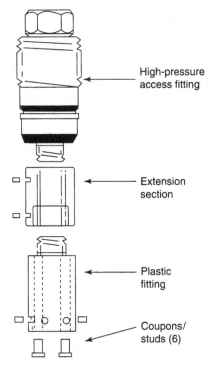

High-pressure
access fitting

Extension
section

Plastic
fitting

Coupons/
studs (6)

FIGURE 5-1c   Petrolite Bioprobe ™ sampler device with high-pressure access fitting. (Reproduced
with permission from Baker Petrolite Corporation.)

biofilm accumulates. Measurements that monitor either heat-transfer or fluid-friction resistance as an indicator for biofilm accumulation must also include some method for actually characterizing the biofilm. Zisson et al. (1996) monitored changes in heat-transfer resistance in a standby service water system of a nuclear power plant using a side-stream device. They concluded that heat-transfer resistance was a more accurate indicator of system fouling than culturing planktonic microorganisms, corrosion-coupon analyses, scale/inhibitor residuals, or differential pressure measurements. Heat-transfer resistance was used to evaluate biocides that would prevent MIC, but the measurements were not directly related to corrosion.

ZERO RESISTANCE AMMETER

Zero resistance ammeter (ZRA) measurements have been used for many years to monitor the electrochemical impact of microorganisms on metal surfaces in laboratory experiments (Gerchakov et al., 1986). Recently, this technique has been used to measure localized corrosion events under field conditions using one of two basic designs: (1) a small-surface-area anode and large cathode, and (2) an occluded cell mimicking small crevices.

One commercially available on-line monitoring device for biofilm development uses an electrode made up of a series of identical UNS 30403 stainless-steel discs (Figure 5-2) (Licina and Carney, 1999). One set of these discs is polarized relative to the other for approximately 1 h day^{-1} (over consecutive polarizations), maintaining the same polarity. An increase in the applied current required to achieve the desired potential between the electrode sets is one criterion for identification of biofilm formation. During the unpolarized periods, the electrode sets are connected through a ZRA and generated current is monitored continuously. Increase in generated current also provides evidence of biofilm formation. The device has been used in fire protection systems, emergency service water stands, and essential equipment cooling water systems in nuclear power plants (Figure 5-3a–c). The device provides a warning when the biofilm maintains a certain current so that the system can be cleaned or biocides added. This technique only indicates a corrosion risk. It does not measure any specific properties of biofilms or any parameter related to corrosion. According to the authors, the low level of polarization of the cathode may also encourage microbial colonization. Lee et al. (2005) demonstrated that polarization does not necessarily encourage biofilm formation in natural water or in artificial media and under all

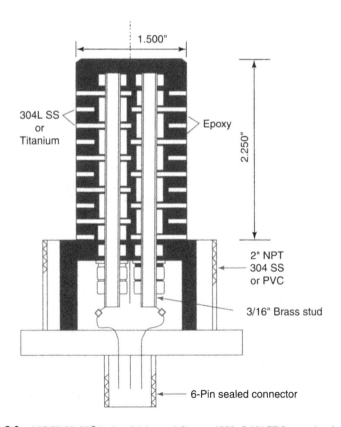

FIGURE 5-2 BIOGEORGE® Probe. (Licina and Carney, 1999. © NACE International, 1999.)

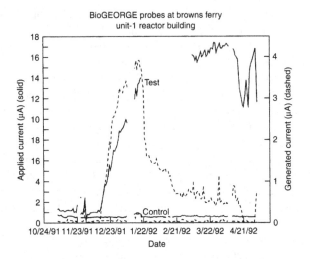

FIGURE 5-3a Biological activity at Browns Ferry Nuclear Plant—fire protection system. (Licina and Carney, 1999. © NACE International, 1999.)

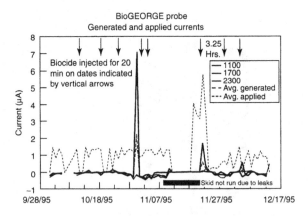

FIGURE 5-3b Biological activity at Susquehanna—emergency service water test stand. (Licina and Carney, 1999. © NACE International, 1999.)

circumstances the biofilms that form after a few hours of exposure to intermittent polarization do not resemble the biofilms that form on unpolarized surfaces in the same medium in terms of areal coverage and microbial composition.

Mollica and Ventura (1993) used a galvanic couple between an undesignated 20Cr-24Ni-6Mo stainless-steel pipe and a 70:30 copper–zinc (UNS C26000) pipe (Figure 5-4) to monitor biofilm growth on surfaces exposed to natural seawater through an integrated monitoring system (Figure 5-5). Results of the field test showed a measurable increase in the galvanic current due to 10^7 cells cm^{-2} (Figure 5-6) and a current decrease after chlorination (Figure 5-7). They concluded

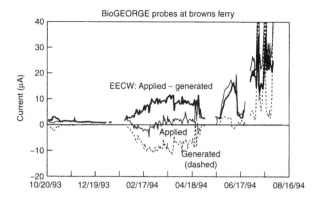

FIGURE 5-3c Biological activity at Browns Ferry Nuclear Plant – essential equipment cooling water. (Licina and Carney, 1999. © NACE International, 1999.)

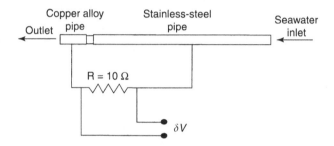

FIGURE 5-4 Schematic of a galvanic loop between undesignated 20Cr-24Ni-6Mo stainless steel and 70/30 copper/zinc (UNS C26000) pipe surfaces exposed to natural flowing seawater. (Mollica and Ventura, 1993. © NACE International, 1993.)

that the device allowed optimization of antifouling treatments by controlling chlorine concentrations and frequency of injections to minimize biofilm recovery rate.

MULTITECHNIQUE APPROACHES

Stokes et al. (1994) described an on-line, real-time fouling and corrosion monitoring system that consisted of a miniature side-stream heat exchanger with an arrangement of corrosion sensors, flow and heat controllers, and a data collection device. Either heat-transfer rate or wall temperature could be controlled. If the heat-transfer rate was controlled, the wall temperature increased as the surface fouled to maintain the set heat rate. If the wall temperature was set, the heating rate decreased to maintain the set wall temperature as the resistance to heat transfer increased. Differences in heat-transfer resistance from that of a clean surface were used as a measure of biofilm formation. Corrosion was monitored using four electrochemical techniques: ZRA, electrochemical

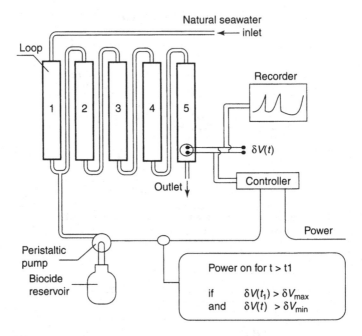

FIGURE 5-5   Schematic of monitoring system for biofilm growth on pipe surfaces exposed to natural flowing seawater. (Mollica and Ventura, 1993. © NACE International, 1993.)

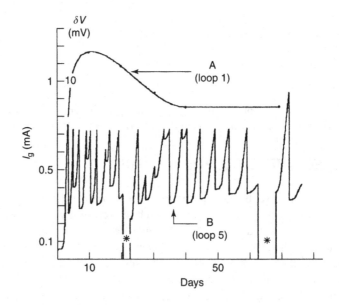

FIGURE 5-6   Trend of the galvanic current in untreated (curve A) and automatically chlorinated (curve B) seawater. *Recorder out of function. (Mollica and Ventura, 1993. © NACE International, 1993.)

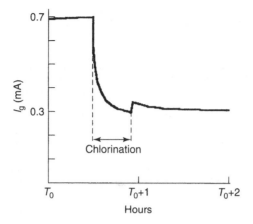

FIGURE 5-7   An example of galvanic current decay during an automatic chlorination. (Mollica and Ventura, 1993. © NACE International, 1993.)

current noise (ECN), electrochemical potential noise (EPN), and linear polarization resistance (LPR). The unit was used for 1 year on a cooling water system of a petrochemical facility that had suffered severe underdeposit corrosion. River water was used in the cooling towers. Figure 5-8 is typical of the differences between the corrosion on a heated surface in a fouled condition and after cleaning. Before cleaning, the values of ECN and ZRA were always near full scale. The trace of EPN showed transients typical of pit initiation. After cleaning, EPN was featureless and the ECN and ZRA values dropped by two and one orders of magnitude, respectively. In contrast, the LPR output indicated a larger active area of apparently increased corrosion activity. During the field trial, corrosion rates as high as 100 milli-inches per year due to underdeposit corrosion were indicated. Metallography confirmed extensive localized corrosion. The authors did not analyze fouling deposits, but based on the use of natural water they assumed biofilm formation.

Enzien and Yang (2001) described a differential flow cell method for monitoring localized corrosion in industrial water (Figure 5-9). In this technique, a combination of LPR and ZRA measurements was used to obtain the rate of localized corrosion for carbon steel (undesignated) in aqueous solutions. The measurements were carried out in an electrolytic flow cell with a large cathode placed in faster flow conditions and two small anodes placed in a slower flow condition. The anodes and cathode were electrically connected via a ZRA. The technique was used in a pilot cooling water test focused on optimizing a scale and corrosion treatment program specifically for localized corrosion (Figures 5-10 and 5-11). They reported that monitoring planktonic bacteria was not effective at predicting microbial fouling or MIC processes. Additionally, general corrosion rates were low throughout experiments; therefore, LPR measurements did not accurately predict a localized corrosion problem.

Uchida et al. (1997) developed an electrochemical device for simulating a single pit (Figure 5-12) consisting of an artificial anode and a carbon steel tube (undesignated) acting as the cathode coupled to a ZRA. The monitoring device was evaluated at a

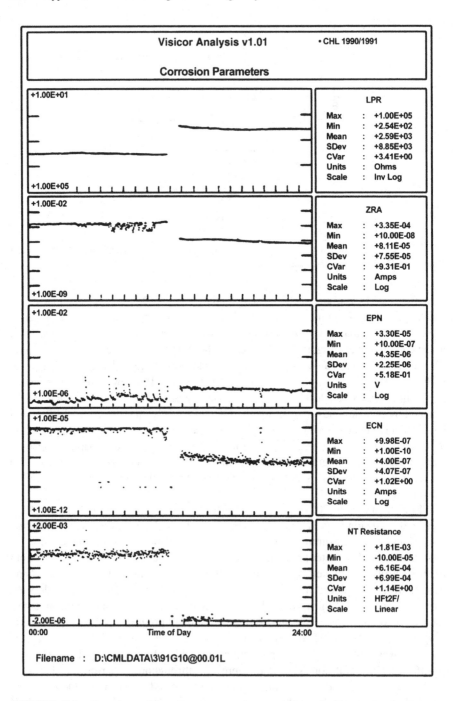

FIGURE 5-8 Corrosion activity on a heated surface in a fouled condition and after cleaning. (Stokes, 1994. Reprinted with permission from STP 1232-Microbiologically Influenced Corrosion Testing, © ASTM International. 100 Barr harbor Drive, West Conshohocken, PA 19428.)

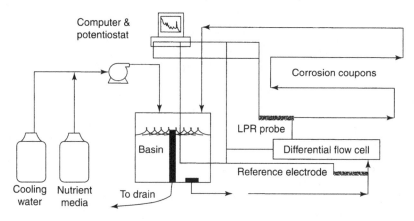

FIGURE 5-9   Intermediate stage unit (ISU) design used for bench studies of MIC monitoring. The ISU used a differential flow cell. (Reprinted from Enzien and Yang, 2001, with permission from Taylor and Francis Ltd. http://www.tandf.uk/journals.)

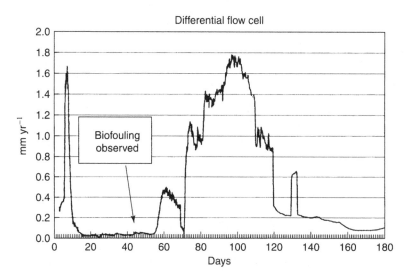

FIGURE 5-10   Estimated localized corrosion rates calculated from ZRA current density measurements in the differential flow cell. (Reprinted from Enzien and Yang, 2001, with permission from Taylor and Francis Ltd. http://www.tandf.uk/journals.)

refinery plant cooling tower. Pitting associated with biofilms had been observed on some heat exchangers during maintenance shutdowns. The goal was to improve the cooling water treatment program by reducing total maintenance costs. Figure 5-13 shows the measured galvanic current versus time for the intermittent dose of biocide treatment (broken line) and the continuous feed of biocide (solid line). In case of intermittent dose, the galvanic current sharply increased up to about 15 µA at the beginning

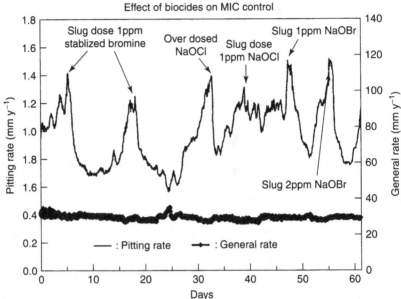

FIGURE 5-11 Pitting and general corrosion rates for the pilot cooling water test. NaOCl was fed continuously at a level of 0.1 ppm free residual halogen; accidental over dosing is indicated with an arrow. Several different treatments are indicated on the graph and were in addition to the continious NaOCl feed. All treatments were slug doses in units of ppm as Cl$_2$. (Reprinted from Enzien and Yang, 2001, with permission from Taylor and Francis Ltd. http://www.tandf.uk/journals.)

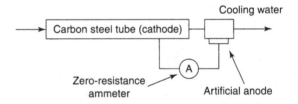

FIGURE 5-12 Schematic diagram of the device developed for monitoring pitting corrosion. (Uchida et al., 1997. © NACE International, 1997.)

and gradually increased thereafter with daily peaks. Finally the current reached about 26 µA at the end of the test period. The current increase was interpreted as poor micro-biological control with intermittent biocide feed. The galvanic current increased at the beginning and gradually decreased to a few microamps in the case of continuous feed. The authors reported that the galvanic current measurement was a more sensitive indicator of biofilm formation and biocide effectiveness for real-time monitoring than pressure drop measurement (Figure 5-14). Iimura et al. (1996) reported that they had succeeded in predicting the penetration rate of carbon steel (UNS G10120 [JIS SS 400]) heat exchanger tubes with a growth model of pitting corrosion and a similar device.

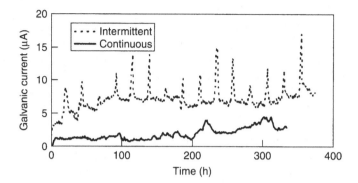

FIGURE 5-13   Change in galvanic current versus time for intermittent chlorination and continuous chlorination. (Uchida et al., 1997. © NACE International, 1997.)

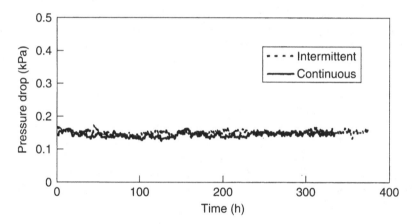

FIGURE 5-14   Change in pressure drop versus time for intermittent chlorination and continous chlorination. (Uchida et al., 1997. © NACE International, 1997.)

Smart et al. (1996, 1997) described an on-line internal corrosion and bacteria monitoring system used in a ship-unloading terminal that received crude oil and other hydrocarbons from tankers or barges. The continuously circulated bypass unit took an oil–water suspension from a pipeline (undesignated), separated the water for testing, and returned the oil to the pipeline. The instrumentation measured the corrosion rate and changes in corrosion characteristics of a pipeline in real-time using multiple techniques including electrical resistance probes, LPR, galvanic probes, hydrogen probes, ac impedance spectroscopy, ECN, pH, and conductivity measurements. The water was used to measure corrosion inhibitor residuals, dissolved iron, bacterial counts, or other measures of bacteria. The system included carbon steel (undesignated) coupons, which could be removed and examined. Analysis of the water chemistry and microbiology as well as the examination of the coupons were off-line and not in real-time. The authors established that rapid MIC was occurring in the unloading pipeline and that the corrosion inhibitor was ineffective. They were able to control the MIC by establishing an effective biocide treatment program.

Videla et al. (1994) described a monitoring program for assessing biodeterioration on undesignated carbon steel and stainless steel in recirculating cooling water systems. The program was based on (1) water quality control, (2) corrosion monitoring in the field (weight loss and linear polarization probe), (3) laboratory corrosion tests (polarization techniques and corrosion potential vs. time measurements), and (4) use of a side-stream sampling device that allowed monitoring of sessile populations, biofilms, corrosion morphology and intensity, and biological and inorganic deposits analysis. Comparison of the corrosive attack on carbon steel coupons maintained with and without biocide indicated that there was little metal attack in the biocide-treated cooling system. Conversely, carbon steel coupons exposed to water without biocide treatment resulted in significant corrosion. They used the multiple monitoring program in an oil field in Mendoza, Argentina, to implement an effective biocide treatment.

Kane and Campbell (2004) described a technique for monitoring real-time MIC using LPR, harmonic distortion analysis (HAD) for measurements of B value (Stearn–Geary constant) for the correction of LPR corrosion rates, and ECN for evaluation of pitting tendencies. Measurements were made in seawater using a three-electrode probe arrangement with three identical carbon steel (UNS G10180) electrodes placed in a test cell attached to a seawater loop, designed to simulate water injection conditions, to which sulfate-reducing bacteria (SRB) were added (Figure 5-15). They demonstrated that the ability to measure the actual B value in the environment resulted in the determination of more accurate corrosion rates in contrast to conventional corrosion monitoring techniques, which use a default B value that does not relate to actual values in the service environment (Figure 5-16). Their study indicated that corrosion rates did not correlate with H_2S production by SRB.

Brossia and Yang (2003) developed a multielectrode array sensor system (MASS) to monitor corrosion in both laboratory tests and industrial processes (Figures 5-17 and 5-18). The probe was used to conduct a series of biotic and abiotic

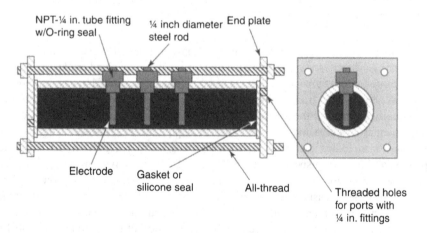

FIGURE 5-15 Schematic of three-electrode probe arrangement used for automated electrochemical monitoring. (Kane and Cambell, 2004. © NACE International, 2004.)

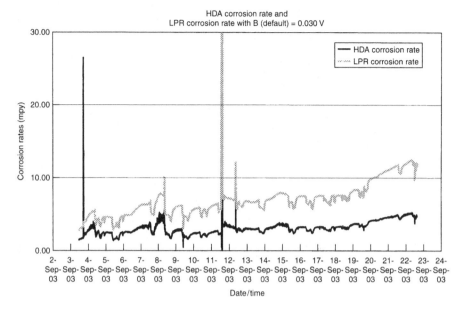

FIGURE 5-16 Comparison between conventional LPR ($B = 0.030\text{V}$) and corrosion rates determined by HAD techniques with on-line B value. (Kane and Campbell, 2004. © NACE International, 2004.)

FIGURE 5-17 Different configurations of MASS probes. (Brossia and Yang, 2003. © NACE International, 2003.)

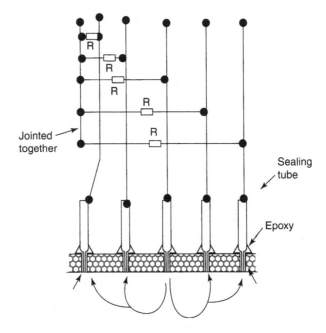

FIGURE 5-18 Schematic of MASS probe. (Brossia and Yang, 2003. © NACE International, 2003.)

tests to determine if the probe could detect MIC. The MASS probe consists of multiple miniature electrodes made of the metals to be studied. The miniature electrodes were coupled together by connecting each of them to a common joint through independent resistors, with each electrode simulating an area of corroding metal. The standard deviation of the currents from the different miniature electrodes was used as an indicator for localized corrosion. Using carbon steel (UNS G10100) electrodes, they were able to demonstrate that corrosion rates increased by an order of magnitude in the presence of SRB compared to sterile controls. For comparative purposes, several test cells were constructed using flat coupons and monitored using LPR to determine corrosion rates under conditions similar to those in the MASS probes. The corrosion rates obtained using the probe were much higher than those determined using LPR. The probe, however, cannot distinguish between biotic and abiotic pitting.

ELECTROCHEMICAL NOISE ANALYSIS

Electrochemical noise analysis (ENA) has been used for monitoring MIC. King et al. (1986) interpreted noise measurements for grey and ductile iron pipes (undesignated) in environments containing SRB as being indicative of film formation and breakdown. Higher noise levels and greater fluctuations indicate localized corrosion. The magnitude of noise fluctuations depends on the total impedance of the system.

A corroding metal undergoing uniform corrosion with fairly high corrosion rates might be less noisy than a passive metal showing occasional bursts of noise due to localized breakdown of the film followed by rapid repassivation.

Iverson et al. (1986) used ENA to monitor corrosion of carbon steel (UNS G10200) in a seawater culture of marine SRB and concluded that breakdown of the iron sulfide film was accompanied by generation of electrochemical noise. Moosavi et al. (1986) presented data collected at E_{corr} for a carbon steel (undesignated) reinforced concrete block exposed to a marine medium containing active SRB in two noise plots: a time record and a potential distribution chart showing the population and the magnitude of the potential fluctuations. Fluctuation in the noise record for 218 days was interpreted as due to the sudden rupture of the protective oxide film followed by immediate repassivation. The events recorded for an exposed rebar showed fluctuations lower by a factor of 10 than those observed with a covered rebar.

Figure 5-19a presents potential and current noise plots for carbon steel (undesignated) in two cooling waters of different salt concentrations (Legat, 1993). Cooling water 2 contains higher concentrations of chlorides and sulfates, producing very different fluctuation patterns. Figure 5-19b shows corresponding power-spectral density (PSD) analyses for the same cooling waters. A higher slope for the spectral curves is obtained in the locally corroding water #2 compared with the uniformly corroding water #1.

Little et al. (1997) used ENA for remote on-line monitoring of carbon steel (undesignated) electrodes in a test loop of a surge water tank at a gas storage field. The experimental design and system for remote ENA and electrochemical impedance spectroscopy (EIS) (discussed later) data collection have been presented elsewhere (Xiao et al., 1997). Noise measurements were compared to electrode weight-loss measurements. Noise resistance (R_n) was defined as

$$R_n = \sigma\{V(t)\}/\sigma\{I(t)\} \tag{5-1}$$

where $\sigma\{V(t)\}$ is the standard deviation of potential fluctuations and $\sigma\{I(t)\}$ the standard deviation of current fluctuations. If one assumes that R_n is close to R_p (polarization resistance), then the instant corrosion current i_c can be obtained from the R_n data as follows:

$$i_c = B/R_n \tag{5-2}$$

where $B = b_a b_c / 2.3(b_a + b_c)$; b_a and b_c are the anodic and cathodic Tafel slopes. ΣINT in grams can be calculated by integration of i_c over the total exposure time t_f:

$$\Sigma INT = C_g \int i_c dt \tag{5-3}$$

where $C_g = 4.7 \times 10^5$ g C^{-1} is used for the conversion of μA h^{-1} for electrodes with a 5.5-cm^2 exposed area. Representative noise data for $\sigma\{V(t)\}$, $\sigma\{I(t)\}$, R_n, and ΣINT are presented in Figure 5-20a–d (Little et al., 1997). The slope of the ΣINT –time curve was steep during the initial exposure, indicating a high corrosion rate that

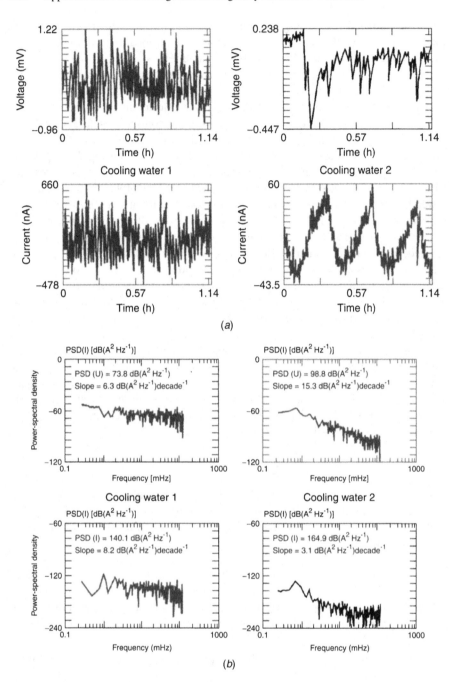

FIGURE 5-19a,b (a) Potential and current noise for carbon steel (undesignated) in different cooling waters and (b) PSD for potential and current noise for carbon steel in different cooling waters. (Legat, 1993. © NACE International 1993.)

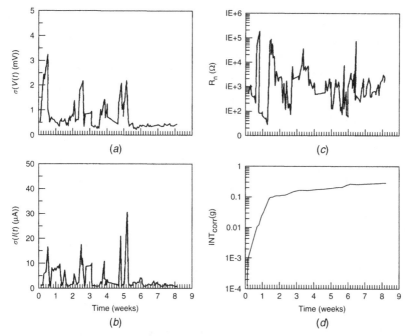

FIGURE 5-20a–d Representative noise data as monitored for carbon steel (undesignated) electrodes exposed in a test loop at a gas storage field: (a) $\sigma\{V(t)\}$, (b) $\sigma\{I(t)\}$, (c) R_n, and (d) INT$_{Corr}$. (Little et al., 1997.)

decreased significantly. The average corrosion rate CR_{int} in $\mu m\ yr^{-1}$ during the exposure period was calculated as follows:

$$CR_{int} = (\sum INT/t_f)C_p \qquad (5\text{-}4)$$

where C_p, a conversion constant, is $8.4 \times 10^4\ \mu m\ g^{-1}$.

$\sum INT$ and t_f are given in grams and years, respectively. CR_{int} values for 19 noise electrodes were compared with the mean corrosion rates determined from weight-loss data (Figure 5-21) (Little et al., 1997). The 19 points were fit to a straight line through the origin using least-squares analysis. Results indicated that the slope of the fit curve was 0.84 with a regression coefficient (R^2) equal to 0.86. The line with slope equal to 1 is for the ideal case, where results obtained using the electrochemical noise data and weight-loss measurements are identical.

Experimental potential and current fluctuations for polymer-coated steel (undesignated) samples exposed to artificial seawater for 780 days (Mansfeld et al., 1998) are shown in Figures 5-22a and b, respectively, while the corresponding power spectral density (PSD) plots are shown in Figure 5-22c and d. The spectral noise plot (Figure 5-22e), also called a noise impedance plot, contains the impedance spectrum determined in an independent measurement. Figure 5-22e demonstrates excellent agreement between the two types of electrochemical measurements, but also illustrates the limited bandwidth Δf in which valid EN measurements can be

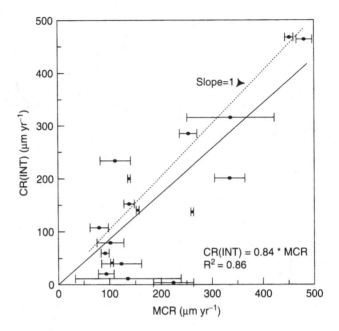

FIGURE 5-21 CR_{int} values for 19 noise probes compared with mean corrosion rates (MCR) determined from weight loss. (Little et al., 1997.)

obtained. Similar results were obtained for polymer-coated steel samples exposed to natural seawater.

R_n is sometimes assumed to be equal to R_p. However, for passive systems such as stainless steel or Ti alloys exposed to neutral media, R_n is several orders of magnitudes lower than R_p. Similar results were reported for polymer-coated steel. In fact, for the epoxy polyamide-coated steel sample, no relationship between R_n and any properties of the coated steel sample could be found. Moreover, R_n was found to depend on Δf (Mansfeld et al., 1998). In general, R_n will be close to R_p only in those cases where impedance is independent of frequency in the bandwidth of the EN measurement.

Little et al. (1999) used ENA to examine spatial relationships between marine bacteria and localized corrosion on polymer-coated steel (UNS G41400). Samples containing intentional defects in polymer coatings were exposed for 30 days in artificial and natural seawater with and without attached zinc coupons providing cathodic protection to the exposed areas. R_n and R_{sn}^o increased with time for all cathodically protected samples due to the formation of calcareous deposits in the defects. Surface analysis showed that very few bacteria were present in the defects of the cathodically protected samples, while large amounts of bacteria were found in the corrosion layers of the freely corroding samples.

ELECTROCHEMICAL IMPEDANCE SPECTROSCOPY

Mansfeld and coworkers used EIS in addition to ENA to monitor coating degradation over an undesignated steel during exposure to natural seawater for periods up

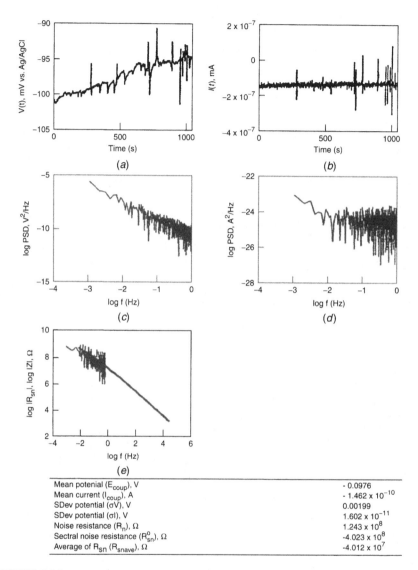

Mean potenial (E_{coup}), V	- 0.0976
Mean current (I_{coup}), A	- 1.462 x 10^{-10}
SDev potential (σV), V	0.00199
SDev potential (σI), V	1.602 x 10^{-11}
Noise resistance (R_n), Ω	1.243 x 10^8
Sectral noise resistance (R_{sn}^o), Ω	-4.023 x 10^8
Average of R_{sn} (R_{snave}), Ω	-4.012 x 10^7

FIGURE 5-22a–e Analysis of electrochemical noise data for an epoxy polyamide coating system exposed to artificial seawater for 780 days: (a) voltage fluctuation with time $V(t)$, (b) current fluctuation with time $I(t)$, (c) voltage PSD versus log of frequency f, (d) current PSD versus log of frequency f, (e) noise impedance plot, log of spectral noise R_{sn} and log of impedance Z versus log of frequency f. (Reprinted from Mansfeld et al., 1998, with permission from Elsevier.)

to 3 years (Mansfeld et al., 1998; Xiao et al., 1997). Figure 5-23a–d is an example of the impedance spectra over a 7-month period. Coating thickness was about 200 μm for all samples. Impedance spectra show a gradual decrease of the impedance modulus |Z| with time for all coatings except for the all-epoxy-polyamide system. Electrochemical impedance spectroscopy data for polymer-coated steel (undesignated) are usually

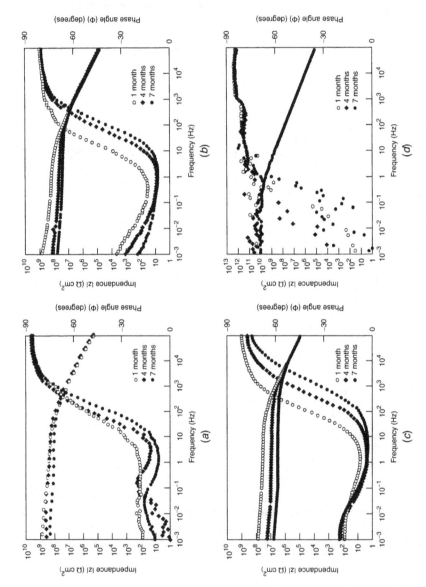

FIGURE 5-23a–d Impedance spectra for coated steel (undesignated) exposed to natural seawater for 1, 4, and 7 months at Port Hueneme, CA: (*a*) zinc primer, epoxy polyamide midcoat, and urethane topcoat; (*b*) zinc primer, epoxy polyamide midcoat, and latex topcoat; (*c*) epoxy polyamide primer and midcoat and latex topcoat; (*d*) epoxy polyamide primer, midcoat, and topcoat. (Reprinted from Mansfeld et al., 1998, with permission from Elsevier.)

fitted to the equivalent circuit shown in Figure 5-24a (Mansfeld, 1995), where C_c is the coating capacitance, R_{po} the pore resistance, C_{dl} the capacitance of the delaminated area A_d, and R_p the corresponding polarization resistance. Spectra for thick coatings should be fit to the open-boundary finite length diffusion (OFLD) model (Figure 5-24b) (Mansfeld et al., 1997, 1998) in which C_c is replaced by a constant-phase element (CPE). The OFLD element is given by

$$Z_{OFLD} = \frac{\tanh(B(j\omega)^{1/2})}{Y_o(j\omega)^{1/2}} \tag{5-5}$$

where $B = l/(D)^{1/2}$ is the characteristic diffusion parameter, l the diffusion length, D the diffusion coefficient, and $Y_o = (\sigma(2)^{1/2})^{-1}$. For infinite values of l, this model becomes the Randles circuit in which the Warburg impedance Z_w is in series with R_p and is given by

$$Z_w = \sigma(1-j)\omega^{-1/2} \tag{5-6}$$

Deterioration of protective coating properties can be followed by determination of A_d estimated from experimental values of R_{po}, C_{dl}, or R_p (Mansfeld, 1995). In corrosion monitoring, it is often not necessary to obtain a full analysis of all impedance spectra and a simpler approach such as the break-point frequency method can be used. A_d can be estimated from the break-point frequency f_b defined as

$$f_b = (2\pi R_{po}C_c)^{-1} = A_d(2\pi R_{po}^o C_c)^{-1} = \Delta(2\pi R_{po}^o C_c^o)^{-1} \tag{5-7}$$

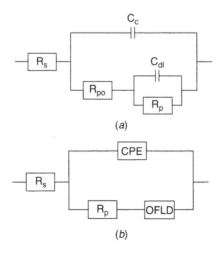

(a)

(b)

FIGURE 5-24a, b   Equivalent circuits used to fit EIS data for polymer-coated steel (undesignated) exposed to seawater: (a) thin coatings (adapted from Mansfeld, 1995); (b) thick coatings, OFLD model. (Reprinted from Mansfeld et al., 1998, with permission from Elsevier.)

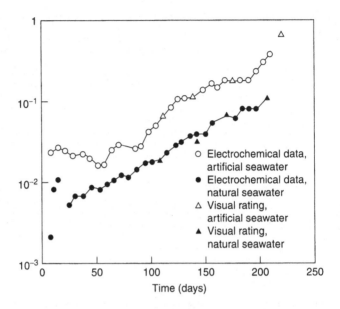

FIGURE 5-25   Time dependence of corroding area obtained from f_b and by visual observation according to ASTM D 610 for a coated steel (undesignated) sample exposed at Port Hueneme, CA. (Reprinted from Mansfeld et al., 1997, 1998, with permission from Elsevier.)

where R^o_{po} is the specific pore resistance ($\pi\ \mathrm{cm}^2$) and C^o_c the specific coating capacitance. $\Delta = A_d/A_t$, where A_t is the total exposed area and A_d the damage function. Figure 5-25 (Mansfeld et al., 1998) gives a comparison of Δ values calculated based on Eq. (5-7). The f_b data were converted into Δ values using the results of visual observation according to ASTM D 610 after about 110 days.

SUMMARY

The monitoring techniques described in this chapter are useful for specific limited applications. All the techniques are based on assumptions that can only be validated by a thorough understanding of the system that one is attempting to monitor. Electrical resistance is appropriate for indicating a change in the general corrosion rate, but the results are difficult to interpret for localized corrosion. Polarization resistance will indicate that something is happening, but will not give an accurate measure of the localized corrosion rate. Electrochemical impedance spectroscopy requires the area of attack to be determined to calculate the corrosion rate. Electrochemical noise techniques have failed to routinely predict MIC events accurately under field conditions. Zero resistance ammeter measurements of galvanic current cannot be directly converted into corrosion rate. The major limitation for MIC-monitoring techniques is the inability to relate microbiology to corrosion in real time.

REFERENCES

Block KP, DiFranco P (1995). Preventing MIC through experimental, on-line fouling monitor. *CORRROSION/1995*, Paper No. 257. Houston, TX: NACE International.

Brossia CS, Yang L (2003). Studies of microbiologically influenced corrosion using a coupled multielectrode array sensor. *CORROSION/2003*, Paper No. 03575. Houston, TX: NACE International, p. 8.

Characklis WG (1981). Microbial fouling: a process analysis. In: Somerscales EFC, Knudsen JG (eds) *Fouling of Heat Transfer Equipment*. Washington, DC: Hemisphere, pp. 251–291.

Enzien M, Yang B (2001). Effective use of monitoring techniques for use in detecting and controlling MIC in cooling water systems. *Biofouling*, **17**(1): 47–57.

Gerchakov SM, Little BJ, Wagner P (1986). Probing microbiologically induced corrosion. *Corrosion*, **42**: 689–692.

Iimura A, Takahashi K, Uchida T (1996). Growth model and on-line measurement of pitting corrosion of carbon steel. *CORROSION/1996*, Paper No. 343. Houston, TX: NACE International.

Iverson WP, Olson GJ, Heverly LF (1986). The role of phosphorous and hydrogen sulfide in the anaerobic corrosion of iron and the possible detection of this corrosion by an electrochemical noise technique. In: Dexter SC (ed) *Biologically Induced Corrosion*. Houston, TX: NACE International, pp. 154–161.

Kane RD, Campbell S (2004). Real-time corrosion monitoring of steel influenced by microbial activity (SRB) in simulated seawater injection environments. *CORROSION/2004*, Paper No. 04579. Houston, TX: NACE International, p. 15.

King RA, Skerry BS, Moore DCA, Stott JFD, Dawson JL (1986). Corrosion behaviour of ductile and grey iron pipes in environments containing sulphate-reducing bacteria. In: Dexter SC (ed) *Biologically Induced Corrosion*. Houston, TX: NACE International, pp. 83–91.

Knudsen JG (1981). Apparatus and techniques for measurement of fouling of heat transfer surfaces. In: Somerscales EFC, Knudsen JG (eds) *Fouling of Heat Transfer Equipment*. Washington, DC: Hemisphere, pp. 57–81.

Lee JS, Ray RI, Little BJ, DeMarco DR, Dorsey MH (2005). An evaluation of an inline sensor for detection of microbial activity. *CORROSION/2005*, Paper No. 05497. Houston, TX: NACE International.

Legat A (1993). Electrochemical noise as the basis of corrosion monitoring. *12th International Corrosion Congress*, Houston, TX: NACE International, pp. 1410-1419.

Licina GJ, Carney CS (1999). Monitoring biofilm formation and incipient MIC in real time. *CORROSION/1999*, Paper No. 175. Houston, TX: NACE International.

Little BJ, Ray RI, Wagner PA, Jones-Meehan J, Lee CC, Mansfeld F (1999). Spatial relationships between marine bacteria and localized corrosion on polymer coated steel. *Biofouling*, **13**: 301–321.

Little BJ, Wagner PA, Ray RI (1997). New experimental techniques in the study of MIC. In: Advanced Monitoring and Analytical Techniques, *Proceedings of Corrosion/97 Research Topical Symposium - Part I*, Houston, TX: NACE International, pp. 31–52.

Mansfeld F (1995). Use of electrochemical impedance spectroscopy for the study of corrosion protection by polymer coatings. *J. Appl. Electrochem.*, **25**: 187–202.

Mansfeld F, Han LT, Lee CC, Chen C, Zhang G, Xiao H (1997). Analysis of electrochemical impedance and noise data for polymer coated metals. *Corros. Sci.*, **39**(2): 255–279.

Mansfeld F, Han LT, Lee CC, Zhang G (1998). Evaluation of corrosion protection by polymer coatings using electrochemical impedance spectroscopy and noise analysis. *Electrochim. Acta*, **43**: 2933–2945.

Mollica A, Ventura G (1993). Use of a biofilm electrochemical monitoring device for an automatic application of antifouling procedures in seawater. In: *Proceedings of the 12th International Corrosion Congress*. Houston TX: NACE International, pp. 3807–3812.

Moosavi AN, Dawson JL, Houghton CJ, King RA (1986). The effect of sulphate-reducing bacteria on the corrosion of reinforced concrete. In: Dexter SC (ed) *Biologically Induced Corrosion*. Houston, TX: NACE International pp. 291–308.

Smart JS, Pickthall T, Wright TG (1996). Field experiences in on-line bacteria monitoring. *CORROSION/1996*, Paper No. 279. Houston, TX: NACE International.

Smart JS, Pickthall T, Carlile A (1997). Using on–line monitoring to solve bacteria corrosion problems in the field. *CORROSION/1997*, Paper No. 212. Houston, TX: NACE International.

Stokes PSN, Winters MA, Zuniga PO, Schlottenmier DJ (1994). Developments in on-line fouling and corrosion surveillance. In: Kerns JR, Little BJ (eds) *Microbiologically Influenced Corrosion Testing*, ASTM Special Technical Publication STP 1232. Philadelphia, PA: American Society for Testing and Materials, pp. 99–107.

Uchida T, Umino T, Arai K (1997). New monitoring system for microbiological control effectiveness on pitting corrosion of carbon steel. *CORROSION/1997*, Paper No. 408, Houston, TX: NACE International.

Videla HA, Bianchi F, Freitas MMS, Canales CG, Wilkes JF (1994). Monitoring biocorrosion and biofilms in industrial waters: a practical approach. In: Kerns JR, Little BJ (eds) *Microbiologically Influenced Corrosion Testing*, ASTM Special Technical Publication STP 1232. Philadelphia, PA: American Society for Testing and Materials, p. 128.

Videla HA, de Mele MFL, Silva RA, Bianchi F, Gonzalez CC (1990). A practical approach to the study of the interaction between biofouling and passive layers on mild steel and stainless steel in cooling water. *CORROSION/1990*, Paper No. 124, Houston, TX: NACE International.

Xiao H, Han LT, Lee CC, Mansfeld F (1997). Collection of electrochemical impedance and noise data for polymer-coated steel from remote test sites. *Corrosion*, 53(5): 412–422.

Zisson PS, Whitaker JM, Neilson HL, Mayne LL (1996). Monitoring and controlling microbiological growth in a standby service water system. *Mater. Perform.*, 35(3): 53–57.

Zelver N, Roe FL, Characklis WG (1985). Potential for monitoring fouling in the food industry. In: Lund D, Plett E, Sandu C (eds) *Fouling and Cleaning in the Food Industry*. Madison, WI: University of Wisconsin.

Chapter 6

Impact of Alloying Elements to Susceptibility of Microbiologically Influenced Corrosion

INTRODUCTION

Commercially pure metals may contain a variety of impurities and imperfections that influence all types of corrosion, including microbiologically influenced corrosion (MIC). For some alloys, such as aluminum, as purity increases the tendency for the metal to corrode reduces proportionately. However, high-purity metals frequently have low mechanical strength, leading to the use of alloying elements to improve mechanical, physical, and fabrication characteristics. Nearly all metals and alloys exhibit a crystalline structure. If an alloying element is added to a base metal, the crystal structure may remain essentially stable and produce a solid solution or single phase. Two or more phase structures can result in some alloy combinations. More than one phase in an alloy usually produces poorer corrosion resistance. Alloying elements affect susceptibility to MIC by providing a corrosion-resistant single phase and by affecting reaction kinetics. However, in practical cases, the primary effect of alloying elements is to stabilize a protective film either mechanically or chemically. Alloying elements also alter the formation, chemical composition, thickness, and tenacity of corrosion products and may increase or decrease susceptibility to MIC.

Welding, a nonequilibrium process, produces nonequilibrium microstructure, altering the size, shape, amount, composition, and distribution of microstructural constituents in the fusion zone and the heat-affected zone (HAZ). Preferential

Microbiologically Influenced Corrosion By Brenda J. Little and Jason S. Lee
Published 2007 by John Wiley & Sons, Inc.

corrosive attack at weld-modified regions has been demonstrated for 304L (UNS S30403), 316L (UNS S31603), AL6XN (UNS N08367) stainless steels, 30/70 copper-nickel alloy 400 (UNS N04400), and aluminum alloy 5086 (UNS A95086). Welding may influence MIC by changing the following (Walsh et al., 1995a):

Surface texture   Welding provides a roughened surface and increases the specific surface area of weld regions.

Solute distribution   The molten metal in a weld does not freeze into a homogeneous alloy.

Phase boundaries   A more extensive interface increases the probability of microbial contact.

Grain size effects   Portions of the HAZ are raised to annealing temperatures. Grains may grow and recrystallization may occur.

Localized melting   Regions of the HAZ may experience transient melting based on constitutional liquation, or on segregation effects. Such liquefaction causes extensive networks of solute-rich material.

Precipitates and inclusions   Undesirable phases may form in the HAZ.

Surface oxidation   The surface of a weld is often exposed to air before it has cooled. The surface oxidizes with a nonpassivating coating.

Residual stress   The severe thermal cycle leaves stresses in the weld region. These stresses can be near yield point.

LOW ALLOY STEEL

Walsh et al. (1995b) demonstrated extensive subgrain boundaries coupled with solute redistribution in the fusion zone as a result of welding and MIC, and suggested that welding created a continuous network of sulfur-rich compounds that increased the sensitivity of carbon steel to MIC. Willis and Walsh (1995) evaluated the effect of minor element content (Si, S, and Ce) (Table 6-1) on the susceptibility of low alloy steels (Table 6-2) to MIC in anaerobic aqueous systems (Tables 6-1 and 6-2). They found that sulfide inclusion sites were strongly correlated with increased MIC susceptibility. Pit initiation and propagation occurred most often at sulfide inclusions. For exposures over 4 h, microbial congregation and proliferation were markedly affected by inclusion size, shape, and composition. Sulfur content was directly correlated with the initiation of pits at inclusion sites and directly related to the number of tubercles that developed a much greater surface roughness. Larger, elongated inclusions increased the probability of pit initiation. Addition of Ce decreased sulfide inclusion length and aspect ratios regardless of other compositional variables. Pit initiation, pit depth, and pit volume decreased when Ce was added to carbon steel.

TABLE 6-1 Minor Alloying Elements of Low Alloy Steels

	Element		
Heat	Si	S	Ce
3037 (000)	0.13	0.007	N/A
3038 (100)	0.47	0.007	N/A
3039 (010)	0.14	0.023	N/A
3040 (110)	0.49	0.020	N/A
3043 (001)	0.14	0.004	0.010
3044 (101)	0.48	0.003	0.010
3045 (011)	0.15	0.018	0.010
3046 (111)	0.48	0.017	0.015

Source: Willis and Walsh (1995. © NACE International).

TABLE 6-2 Alloy Composition of Low Carbon Steel (wt.%)

Element				
C	Cr	Mn	Mo	Ni
0.32	0.50	0.80	0.20	0.55

Source: Willis and Walsh (1995. © NACE International).

COPPER AND NICKEL ALLOYS

Copper alloys are frequently used for seawater piping systems and heat exchangers due to their good corrosion resistance combined with mechanical workability, excellent electrical and thermal conductivity, ease of soldering and brazing, and resistance to macrofouling. Alloying nickel and small amounts of iron into copper results in a single-phase structure and increases resistance to turbulence-induced corrosion. Additions of nickel and iron improve the mechanical properties of copper alloys but may increase susceptibility to MIC. In oxygenated seawater, a film of cuprous oxide, cuprite (Cu_2O), forms on predominately copper alloys. Copper ions and electrons pass through the film. Copper ions dissolve and precipitate as $Cu_2(OH)_3Cl$, independent of alloy chemistry.

Differential aeration, selective leaching, underdeposit corrosion, and cathodic depolarization have been reported as mechanisms for MIC of copper alloys. Pope et al. (1984) proposed that the following microbial products accelerated localized attack: CO_2, H_2S, NH_3, organic and inorganic acids; metabolites that act as depolarizers; and sulfur compounds such as mercaptans, sulfides, and disulfides.

In the presence of sulfides, copper alloys form a porous layer of cuprous sulfide with the general stoichiometry $Cu_{2-x}S$, $0 < x < 1$. Copper ions migrate through the layer, react with more sulfide, and produce a thick black scale. The McNeil–Odom model (1992) described in Chapter 2 was used to predict sulfide corrosion in products on copper alloys. Negative standard free energies of reactions were used to predict sulfate-reducing bacteria (SRB)-MIC for copper alloys. Analysis of sulfide corrosion products recovered from corroding copper alloys confirmed the prediction. McNeil et al. (1991) reported that sulfide corrosion products on 99 copper (UNS C11100) were consistently nonadherent, while those on 90:10 (UNS C70600) and 70:30 copper–nickel (UNS C71500) were adherent in SRB laboratory cultures and in natural waters.

While the role of the biofilm in copper pitting is not entirely clear, it appears that the presence of the biofilm contributes to corrosion by maintaining enhanced local chloride concentrations and differential aeration cells (Paradies et al., 1990). Pope (1987) documented MIC of C70600 copper–nickel, and undesignated alloys including admiralty and aluminum brass, and welded aluminum bronze at electric generating facilities using fresh or brackish cooling waters. Most of the C70600 tubes had underdeposit corrosion due to the formation of deposits by slime-forming organisms in association with iron- and manganese-depositing bacteria. Ammonia-producing bacteria were isolated from scale and organic material on the admiralty brass tubes suffering ammonia-induced stress corrosion cracking. Mansfeld and Little (1992) reported that five copper alloys [C11100, C70600, C71500, C44300 (admiralty bronze), and C61400 (aluminum brass)] exposed to natural seawater were colonized by bacteria within 3 weeks, independent of alloy composition. Corrosion rates were higher in natural seawater compared with artificial seawater for all copper alloys exposed. Little et al. (1988) demonstrated that porous C70600 welds provided increased sites for colonization compared to smooth pipe surfaces (Figure 6-1).

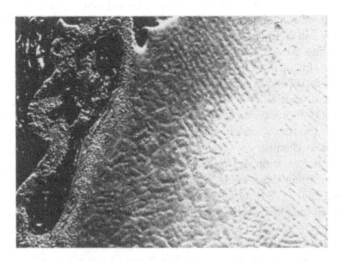

FIGURE 6-1   Cross section of epoxy-embedded 90/10 copper-nickel (UNS C70600) butt weld before exposure, demonstrating weld porosity. (Little et al., 1988.)

Nickel minerals have not been found in corrosion products on C70600 or C71500 copper–nickel alloys. Selective dealloying of zinc, nickel, and iron from copper alloys has been reported by several investigators. Little et al. (1990) demonstrated dealloying of nickel from a C70600 copper–nickel in association with SRB. Mansfeld and Little (1992) and Wagner et al. (1992) described dealloying of nickel in C71500 copper–nickel exposed to flowing natural seawater (Figure 6-2a, b).

Nickel alloys, including 70:30 nickel–copper alloy 400 (N04400) (Table 6-3), are used extensively in highly aerated fast-moving seawater environments as

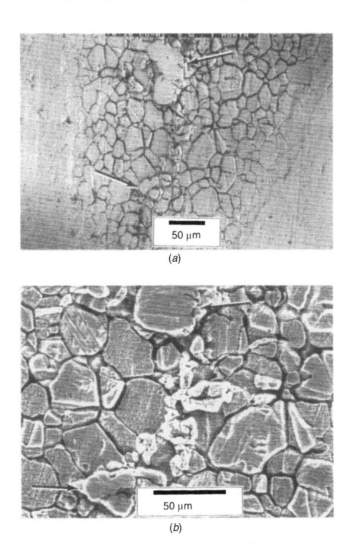

(a)

(b)

FIGURE 6-2 a, b   Pit interiors on 70/30 copper-nickel (UNS C71500) showing intergranular corrosion. Arrows indicate grains of copper. (Mansfeld and Little, 1992.)

TABLE 6-3 Chemical Composition of Nickel-Based Alloys (wt.%)

Alloy (UNS)	Ni	Cu	Cr	Fe	Mn	C	Si	S	Co	Mo	W	Nb	Ti	Al	V
400 (N04400)	63.0 min	28.0–34.0	–	2.5	0.2	0.3	0.5	0.024	–	–	–	–	–	–	–
625 (N06625)	58.0 min	–	20.0–23.0	5.0	0.5	0.1	0.5	–	1.0	8.0–10.0	–	3.15–4.15	0.4	0.4	–
22 (N06022)	Bal.	–	20.0–22.5	2.0–6.0	0.5	0.015	0.08	–	2.5	12.5–14.5	2.5–3.5	–	–	–	0.35

Source: Davis (1998).

evaporators, heat exchanger pumps and valves, diffusers for steam nozzles in steam ejectors, and turbine blades. Uhlig and Revie (1985) calculated that a critical nickel concentration of 35 percent was required for passivity. Copper–nickel alloys containing less than this amount behave like copper. The formation of the protective film on nickel is aided by the presence of iron, aluminum, and silicon. In high-velocity seawater, nickel alloys are superior to predominantly copper alloys because the protective surface film remains intact under highly turbulent and erosive conditions.

Nickel–copper alloys are susceptible to MIC. For example, N04400 is susceptible to pitting and crevice corrosion attack where chlorides penetrate the passive film. Sulfides can cause either a modification or breakdown of the oxide layer. Schumacher (1979) reported that N04400 was susceptible to underdeposit corrosion and oxygen concentration cells formed by bacteria. Gouda et al. (1993) demonstrated pitting of N04400 tubes exposed in Arabian Gulf seawater where pits developed under deposits of SRB and nickel was selectively dealloyed. Little et al. (1990) reported selective dealloying in N04400 in the presence of SRB from an estuarine environment. Pitting occurs when passivity breaks down at local sites. In low-flowing or stagnant conditions, pitting or crevice corrosion may develop in N04400 under deposits of bacteria. Pope (1987) reported a case study from nuclear power plants in which severe pitting corrosion associated with dealloying was observed under discrete deposits on N04400 heat exchanger tubes. Deposits formed by iron- and manganese-depositing bacteria in association with SRB contained large amounts of iron and copper, significant amounts of manganese and silicon, and reduced amounts of nickel.

No evidence for MIC in nickel–chromium and nickel–chromium–molybdenum alloys has been reported. Enos and Taylor (1996) demonstrated that SRB did not cause corrosion of welded alloy 625 (UNS N06625) (Table 6-3). Alloy 22 (UNS N06022) (Table 6-3), a potential candidate packaging material for nuclear waste containment, was evaluated in a simulated, saturated repository environment consisting of crushed rock from the repository site and a continual flow of simulated groundwater for periods up to 5 years (Martin et al., 2004). They were able to demonstrate micron-scale surface alterations due to the presence and activities of microorganisms. Culture conditions were optimized with thiosulfate as an energy source to produce a high-density, metabolically active culture throughout a 7-month exposure.

STAINLESS STEELS

The corrosion resistance of stainless steels is due to the formation of a thin passive chromium-iron oxide film at additions of chromium in amounts of 12 percent or more. Metal-depositing organisms, important in MIC of stainless steels, may catalyze the oxidation of metals, accumulate abiotically oxidized metal precipitates, or derive energy by oxidizing metals. Dense deposits of cells and metal ions create oxygen concentration cells that effectively exclude oxygen from the area immediately under the deposit. Underdeposit corrosion is important because it initiates a series of events that are, individually or collectively, extremely corrosive (Shreir, 1994). As described in Chapter 2, in an oxygenated environment, the area immediately under the deposit becomes a relatively small anode compared to the large surrounding cathode. Cathodic reduction of oxygen may result in an increase in pH of the solution in the vicinity of the cathode. The metal will form metal cations at anodic sites. If the metal hydroxide is the thermodynamically stable phase in the solution, metal ions will be hydrolyzed by water and H^+ ions will be formed. If cathodic and anodic sites are separated from one another, the pH at the anode will decrease and that at the cathode will increase (Figure 6-3). The pH within anodic pits depends on specific hydrolysis reactions of the alloying elements (Table 6-4). The resulting corrosion is due to the formation of acidic ferric chlorides.

Kovach and Redmond (1997) suggested positive correlations between the ferric chloride test and service experience of stainless steels, meaning that resistance to MIC could be predicted from alloy composition (Table 6-5). Accordingly, their hypothesis predicts that resistance to MIC would increase with increased concentrations of chromium and molybdenum in an alloy. For example, there are numerous reports of MIC in S30400 and S31600 (Table 6-6). Scott and Davies (1989) reported failure of a 904L (UNS N08904) (Table 6-6) heat exchanger after exposure to stagnant brackish and seawater that had been excessively chlorinated. However, there are no known instances of MIC during actual service of 6 percent molybdenum stainless steel, including AL6XN (UNS N08367), SMO 254 (UNS S31254), and 1925hMo (N08926), in spite of the fact that these steels have been used

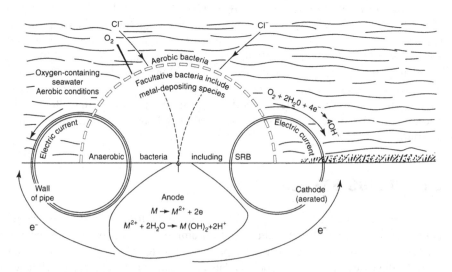

FIGURE 6-3   Reactions possible under tubercles created by metal-depositing bacteria. (Little et al., 1991.)

TABLE 6-4   Specific Hydrolysis Reactions

Hydrolysis reaction	Equilibrium pH
$Fe^{2+}+2H_2O \rightleftharpoons Fe(OH)_2+2H^+$	$pH=6.64-\frac{1}{2}\log a_{Fe^{2+}}$
$Cr^{3+}+3H_2O \rightleftharpoons Cr(OH)_3+3H^+$	$pH=1.53-\frac{1}{3}\log a_{Cr^{3+}}$
$Ni^{2+}+2H_2O \rightleftharpoons Ni(OH)_2+2H^+$	$pH=6.5-\frac{1}{2}\log a_{Ni^{2+}}$
$Mo^{3+}+2H_2O \rightleftharpoons MoO_2+4H^++e^-$	$pH^a=(0.311-0.059\log a_{Mo^{3+}})/0.236$
$Mn^{2+}+2H_2O \rightleftharpoons Mn(OH)_2+2H^+$	$pH=1.53-\frac{1}{3}\log a_{Mn^{2+}}$

$^aE = -0.20$ V (vs. SCE).
Source: Shreir et al. (1994). Reprinted with permission from Elsevier.

TABLE 6-5   Chemical Composition of Stainless-Steel Alloys (wt.%)

Alloy (UNS)	C	Cr	Mn	Si	Ni	P	S	Mo	Cu
304 (S30400)	0.08	18.0–20.0	2.0	1.00	8.0–10.5	0.045	0.03	–	–
304L (S30403)	0.03	18.0–20.0	2.0	1.00	8.0–12.0	0.045	0.03	–	–
308 (S30800)	0.08	19.0–21.0	2.0	1.00	10.0–12.0	0.045	0.03	–	–
316L (S31603)	0.03	16.0–18.0	2.0	1.00	10.0–14.0	0.045	0.03	2.0–3.0	–
410 (S41000)	0.15	11.5–13.5	1.00	1.00	–	0.04	0.03	–	–
904L (N08904)	0.02	19.0–23.0	2.00	1.00	23.0–28.0	0.045	0.035	4.0–5.0	1.0–2.0

Source: Davis (1998).

extensively in service water and other waters known to produce MIC (Ward et al., 1991; Licina et al., 1991; Zhang and Dexter, 1995). Hanford (1990) reported that no MIC has been observed for the 6 percent Mo stainless-steel service water piping system at the Robinson Station of Carolina Power and Light used to replace S30400 that had failed due to MIC. Felder and Stein (1994) performed an extensive field study of the potential of MIC in high-performance steels. They exposed a series of pipe spools in a test loop maintained with lake water at 500 to 600 ppm chloride and pH 7.8 to 8.2 for up to 4 years under four different flow conditions: two flow rates, intermittent flow, and stagnant. Pipe spools were made of S30400, S31600, S31603

TABLE 6-6 Summary of the Literature Reference on MIC and High-Performance Stainless Steel

Duration	Material	Specimen	Environment	Bacterium
14-day service	904L	Heat exchange tubes cooling H_2SO_4	Seawater with sodium hypochlorite	*Desulfotomaculum*
146-day lab	904L	Tube samples	Various cultures in seawater	*Desulfotomaculum* and *Desulfovibrio desulfuricans*

Conclude MIC pitting in 904L based on cavernous pits at tube bottom, possible chlorine contribution investigated and rejected. Pit initiation and propagation in 904L associated with

| 1-year field test | Steel admiralty T304 T316 6Mo stainless high Cr ferritic | Electrodes 600 grit | Catawba Station service water side-stream various flow | No data given |

Steel exhibited constant general corrosion. Admiralty exhibited pitting corrosion. All stainless steels exhibited good resistance to general corrosion. No pitting was exhibited on the stainless steels. A microbiological effect did not seem to be significant

| 14 years | T304 | Service pipe | Soft water | Decayed vegetable matter |
| 18 months | 6Mo stainless | Service pipe | Soft water | Decayed vegetable matter |

Corrosion evaluation is the percentage of metal surface covered by corrosion product (photomicrographs seemed to show evidence of differing colonization but no evidence of corrosion). Changes in concentration of alloying element did not affect "corrosion." "Corrosion" occurred on all alloy welds and HAZs

| 2-day lab | T316 6Mo stainless | Coupons | Electrochemical | Culture of Ni–Cu resistant bacteria from deposits associated with MIC on admiralty brass |

254 SMO and AL6XN had higher corrosion than T316 from OCP and EIS. Any observations of actual pitting, if made, were not reported

| 120-day lab | T304, T316, 904L, 6Mo stainless alloy, 625 | Coupons cut from tube | Exposure to batch culture medium | Eleven different cultures including *Desulfivibrio* and *Thiobacillus ferroxidans* |

All alloys were susceptible to corrosion based on microscopic examination. T316 performed best and 6Mo alloys the worst. (The only evidence of corrosion presented was high SEM micrographs showing a finely pitted surface, with no comparison to the unexposed surface and no evidence of colonization.)

TABLE 6-6   *Continued*

Duration	Material	Specimen	Environment	Bacterium
84-day lab	254 SMO, Ti	Coupons	Electrochemical culture	SOB
167 day lab	254 SMO, Ti	Coupons	Electrochemical culture	SRB
? day lab	Ti	Coupons	Electrochemical culture	H$_2$-producing bacterium
1-year field	254 SMO, Ti	Coupons	Seawater	Biofilm forming

No corrosion current or localized corrosion observed. Calculated corrosion rate of 0.2 mpy at 55 °C and micropits on 254 SMO. No corrosion or hydride formation on titanium. All surfaces remained shiny and free of localized corrosion

15 days	6Mo stainless	Coupons with welds	Electrochemical flowing seawater	Fe-oxidizing, SRO, H production
75 days	Titanium			No data given

Electrochemical data showed very low corrosion for stainless and titanium. No microscopic corrosion of Ti or AL6XN in S/Fe-oxidizing bacteria. Deep etching was observed on 254 SMO in S/Fe-oxidizing bacteria. (Photomicrographs did not show the condition of the Ti and AL6XN surface because they were covered with deposits; the 254 SMO had etch pits about 2 μm in diameter; in a private communication, the author stated the AL6XN showed no pits under the deposit and 254 SMO was free of pits before testing; however, the 254 SMO corrosion was very "shallow".)

6 to 36 weeks	T304L, alloy 800, Sea-Cure, Sanicro 28, AL6XN titanium	Tube samples	16 medium cultures, with NaCl, and sterile controls	SRB

Conclusions are based on microscopic examination of coupons. No measurable weight loss but all alloys corroded even under sterile conditions. The vast majority of corrosion was found on AL6XN. Alloy comparison in order of declining corrosion was AL6XN, Sea-Cure, 304L, Alloy 800, Sanicro 28 = Titanium

4-year field test	T304, T316, 6Mo stainless, titanium	Pipe spools	Natural cooling water, various flow and bicocide, 5/600 ppm CI	Natural colonization, acid-producing and sulfate-reducing *Gallionella*

Severe colonization and pitting in T304 and T316. No colonization or corrosion on titanium. Less than 1% colonization on 6Mo stainless. Two of the 336, 6Mo specimens developed shallow surface attack under a nodule

Source: Kovach and Redmond (1997). Reprinted with permission from Stainless Steel World Magazine (www.stainless-steel-world-net).

and two 6 percent Mo stainless steels (N08367 and N08926). The spools included fittings and a variety of fabrication techniques. Extensive microbial colonization and deep pitting were observed in the S30400 and to a lesser extent in the S31600. In contrast, only 2 of 336 6 percent Mo specimens developed superficial etching under nodules containing bacteria. There are a few laboratory claims of MIC in 6 percent Mo stainless steels. Walsh et al. (1992) and Dowling et al. (1989, 1990) evaluated corrosion of S31254 and N08367 by polarization resistance measurements and surface observation. They concluded that colonization was concomitant with decreased polarization resistance (R_p) and increased corrosion potential (E_{corr}). There was no further evidence of MIC. Little et al. (1992, 1993) demonstrated surface etching of S31254 under two laboratory conditions: exposure to SRB after 167 days at 55 °C and after exposure to sulfur-oxidizing bacteria (SOB) at 25 to 40 °C.

One of the most common forms of MIC attack in austenitic stainless steels is pitting at or adjacent to welds at the heat-affected zone, the fusion line, and in the base metal (Kobrin, 1976). Borenstein (1991) made the following observations for MIC in S30403 and S31603 weldments: Both austenite and delta ferrite phases may be susceptible; combinations of filler and base materials have failed, including matching, higher, and lower alloyed filler combinations; and solution annealing and pickling may produce welds that are less susceptible. A lack of sensitization in austenitic welds did not ensure protection. Additionally, surface conditions commonly associated with corrosion resistance, such as heat tint, and those related to residual stresses, including gouges and scratches, may increase susceptibility. Kearns and Borenstein (1991) state that welds having filler-metal compositions matching the base metal have lower corrosion resistance than fully annealed base metal due to lack of homogeneity and the microsegregation of chromium and molybdenum. Chemically depleted regions can be much more susceptible to localized attack. Stein (1991) reported that MIC susceptibility of base metal related to weld area was not related to sensitization but to the microstructure produced during the manufacturing process of S30400, S31600 and S31603. Reannealing reduced the severity of the pitting corrosion. Videla et al. (1990) observed that sensitization heat treatments with related carbide precipitation lowered pitting corrosion resistance of S30400 and 410 (UNS S41000) stainless steels in the presence of SRB and aggressive anions. Sreekumari et al. (2004) evaluated bacterial attachment to welded S30403 stainless-steel coupons with 308L (UNS S30803) filler metal and the significance of substratum microstructure (Figure 1-8; Chapter 1). Their results demonstrated that bacteria colonized more weld coupons than base-metal coupons. Elemental segregation during welding and/or differential energy distribution between the matrix and the grain boundaries was suggested as the possible reason for the pattern of attachment. Amaya et al. (2002) demonstrated that bacterial adhesion increased at the toe of the weld (Figure 6-4) and heat-affected zone of

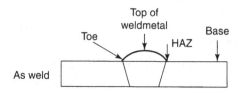

FIGURE 6-4   Schematic illustration of test coupons. (Amaya et al., 2002. © NACE International, 2002.)

S30400 stainless-steel coupons with 308 (UNS S30800) filler in incubations with some stirring (reciprocal shaking). They attributed pitting at the toe of the weld to increased bacterial adhesion and a defective passive film at that location.

Ennoblement of corrosion potential (E_{corr}) for stainless steels during exposure to natural waters has been reported by several investigators. The practical importance of ennoblement is increased probability of localized corrosion as E_{corr} approaches the pitting potential E_{pit}. Johnsen and Bardal (1985) reported that E_{corr} approached -50 mV [vs. saturated calomel electrode (SCE)] after 28 days for stainless steels (Figure 2-1, Chapter 2) that contained 1 to 3 percent molybdenum (UNS S31603 and N08904). In contrast, stainless steels containing 6 percent molybdenum (UNS N08028, S31254, and S44635, and non-UNS designated NU984LN) reached values of $+50$ to 150 mV$_{SCE}$ in the same time period. Early reports suggested that molybdenum was not found in the passive layer for alloys containing up to 5.2 percent molybdenum (Yaniv et al., 1977). However, Olefjord and Wegrelius (1993) studied the influence of molybdenum and nitrogen on the corrosion behavior of high-alloyed stainless steels containing 6 percent molybdenum. Passive layers contained 8 to 4 percent molybdenum.

Buchanan et al. (1996) exposed weldments representative of a range of marine structural materials to the natural marine environment at Delaware Bay. The weldments that exhibited preferential attack included S30403 and S31603, but not N08367.

ALUMINUM AND ALUMINUM ALLOYS

The corrosion resistance of aluminum and its alloys is due to an aluminum oxide passive film. Anodizing produces thicker insulating films and better corrosion resistance. The natural film on aluminum alloys can be attacked locally by halide ions. The susceptibility of aluminum and its alloys to localized corrosion makes it particularly vulnerable to MIC. Most reports of MIC are for 99 percent aluminum alloys (UNS A91xxx group), A92024, and A97075 alloys (Table 6-7) used in aircraft or in underground fuel storage tanks (Salvarezza and Videla, 1984). Localized corrosion attributed to MIC occurs in the water phase of fuel–water mixtures in the bottom of tanks and at the fuel–water interface. Contaminants in fuel include surfactants, water, and water-soluble salts that encourage growth of

TABLE 6-7   Composition Limits for Wrought Aluminum Alloys (wt.%)

Alloy(UNS)	Al	Si	Fe	Cu	Mn	Mg	Cr	Zn	Ti
2024 (A92024)	Bal.	0.50	0.50	3.8–4.9	0.3–0.9	1.2–1.8	0.10	0.25	0.15
7075 (A97075)	Bal.	0.40	0.50	1.2–2.0	0.30	2.1–2.9	0.18–0.28	5.1–6.1	0.20

Source: Davis (1998).

bacteria and fungi. Two mechanisms for MIC of aluminum alloys have been documented: production of water-soluble organic acids by bacteria and fungi, and formation of differential aeration cells. Walsh (1999) demonstrated that bacteria preferential colonized welds and proliferation of bacteria was associated with specific metallurgical features on UNS A92195, an Al-Li-Cu alloy. Specifically, microorganisms aggregated in the depression between dendrite arms in the weld structure. Welding markedly increased the propensity for MIC in all heats; the weld region was equally attacked. The experiments described were conducted at 30 °C.

TITANIUM AND TITANIUM ALLOYS

There are no case histories of MIC for titanium and its alloys. Schutz (1991) reviewed mechanisms for MIC and titanium's corrosion behavior under a broad range of conditions. He concluded that at temperatures below 100 °C, titanium is not vulnerable to iron/sulfur-oxidizing bacteria, SRB, acid-producing bacteria, differential aeration cells, chloride concentration cells, or hydrogen embrittlement. In laboratory studies, Little et al. (1992, 1993) did not observe any corrosion of ASTM grade 2 titanium (UNS R50400) in the presence of SRB or iron/sulfur-oxidizing bacteria at mesophilic (23 °C) or thermophilic (70 °C) temperatures. Using the model of McNeil and Odom (1992) (see Chapter 2), one would predict that titanium would be immune to SRB-induced corrosion. There are no standard free-energy reaction data for the formation of a titanium sulfide. If one assumes a hypothetical sulfide product to be titanium sulfide, the standard enthalpy of reaction is +587 kJ. While standard free energies of reaction are not identical to standard enthalpies of reaction, it is still unlikely that titanium will be derivatized to the sulfide under standard conditions of temperature and pressure. Horn et al. (2001) evaluated the corrosion of ASTM grade 7 titanium (UNS R52400) (Table 6-8) under extreme conditions, exposed to *Thiobacillus ferroxidans*, an SOB. Culture conditions were optimized with thiosulfate as an energy source to produce a high-density, metabolically active culture throughout a 7-month exposure. The surface of R52400 titanium was roughened. Differences with findings by Little et al. (1992, 1993) may be related to alloying elements in the two materials.

TABLE 6-8   Chemical Compositions of Unalloyed Grade Titanium (wt.%)

Grade(UNS)	N	C	H	Fe	O	Pd	Ti
Grade 2 (R50400)	0.03	0.08	0.015	0.03	0.25	—	Bal.
Grade 7 (R52400)	0.03	0.08	0.015	0.03	0.25	0.20	Bal.

Source: Davis (1998).

ANTIMICROBIAL METALS

Sreekumari et al. (2002) evaluated the antimicrobial properties and corrosion behavior of silver-alloyed S30400 stainless steel and its welds. In an effort to make stainless steels antimicrobial and thus enhance its resistance to MIC, two types of stainless steels were prepared: silver-coated and silver-alloyed (Table 6–9). Bacterial adhesion to these materials was compared to adhesion on S30400 stainless-steel control. The area of bacterial adhesion was found to be significantly less in the case of silver-incorporated coupons compared to the controls (Figure 6-5). Silver-alloyed coupons showed more antimicrobial effect than silver-painted coupons. Scanning electron microscopy (SEM) observations indicated that control coupons pitted within 30 days. Corrosion potential showed ennoblement in the case of the control coupons (Figure 6-6*a, b*). Silver-coated coupons also showed an increase in potential while silver-alloyed coupons did not. These experiments were completed with *Pseudomonas* sp. in a dilute nutrient medium.

Nandakumar et al. (2002) evaluated the antimicrobial properties of a magnesium alloy, AZ31B (UNS M11311) (Table 6-10), in laboratory experiments with *Pseudomonas* sp. They observed a high rate of bacterial attachment on days 1 and 2.

TABLE 6-9 Chemical Compositions of Silver Incorporated Stainless Steels

Sample	C	Si	Mn	P	S	Cr	Ni	O	Ag
Silver-alloyed	0.054	0.43	1.0	0.037	0.004	18.50	8.00	0.0062	0.039
Silver-coated	0.042	0.26	1.03	0.035	0.003	18.34	8.30	0.0040	<0.001

Source: Sreekumari et al. (2002. © NACE International).

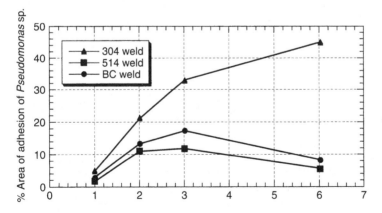

FIGURE 6-5 Variation in percentage of area of bacterial adhesion on UNS S30400 weld coupons as a function of exposure time: 304—control; 514—silver-alloyed; BC—silver-coated. (Sreekumari et al., 2002. © NACE International.)

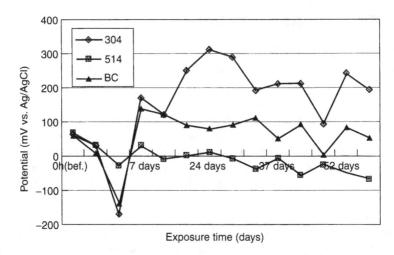

FIGURE 6-6a   Variation in corrosion potential of different UNS S30400 base-metal coupons as a function of exposure time: 304—control; 514—silver-alloyed; BC—silver-coated. (Sreekumari et al., 2002. © NACE International.)

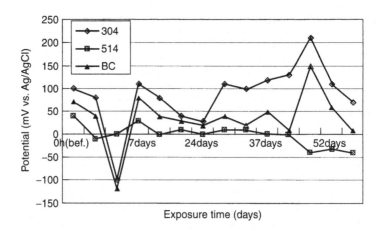

FIGURE 6-6b   Variation in corrosion potential of different UNS S30400 weld coupons as a function of exposure time: 304—control; 514—silver-alloyed; BC—silver-coated. (Sreekumari et al., 2002. © NACE International.)

However, by day 6, the coverage was reduced. The total viable numbers were also reduced (Figure 6-7). The authors determined that magnesium reacts with water to produce magnesium hydroxide, which forms a film on the metal surface. The formation of the film caused an elevation of surface pH into the alkaline region. The combined effect of high pH and concentration of Mg ion adversely affected the growth and survival of *Pseudomonas* sp. both in the medium and on the surface.

TABLE 6-10   Chemical Element Composition (%) of Magnesium Alloy AZ31B

Alloy(UNS)	Al	Mn	Zn	Si	Cu	Ni	Fe	Ca	Mg	Others total
AZ31B (M11311)	3.0	0.46	1.0	<0.01	<0.01	0.001	0.003	<0.01	Remaining portion	0.30

Source: Nandakumar et al. (2002). (Reprinted from Nandakumar, 2002, with permission from Taylor and Francis Ltd. http://www.tandf.uk/journals.)

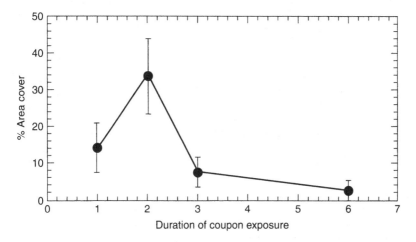

FIGURE 6-7   Variation in percentage of area cover of bacterial adhesion on magnesium alloy (UNS M11311) coupons during the study period (days). (Reprinted from Nandakumar, 2002, with permission from Taylor and Francis Ltd. http://www.tandf.uk/journals.)

They concluded that over the 6-day period, magnesium alloys exhibited antibacterial properties that prevented bacterial attachment and biofilm formation.

SUMMARY

Alloying elements affect susceptibility to MIC by providing a corrosion-resistant single phase and by affecting reaction kinetics. However, in practical cases, the primary effect of alloying elements is to stabilize a protective film either mechanically or chemically. Susceptibility of metals to attack by SRB can be predicted based on the standard free energies of reactions of sulfides with metals. Additions of nickel and iron improve the mechanical properties of copper alloys but may increase susceptibility to SRB-induced MIC. Resistance to MIC can be correlated to increased concentrations of chromium and molybdenum and can be predicted based on resistance to ferric chloride. There are no known instances of MIC in actual service of any stainless steel containing more than 6% molybdenum. Silver and magnesium alloys exhibit antimicrobial properties that could potentially enhance resistance to MIC.

REFERENCES

Amaya H, Miyuki H, Takeishi Y, Ozawa M, Kikuchi Y (2002). Effects of shape of weld bead on bacterial adhesion and MIC occurrence at stainless steel welded joints. *CORROSION/2002*, Paper No. 02556. Houston, TX: NACE International.

Borenstein SW (1991). Microbiologically influenced corrosion of austenitic stainless steel welments. *Mater. Perform.*, **30**(1): 52–54.

Buchanan RA, Kovacs AL, Lundin CD, Khan KK, Danko JC, Angell P, Dexter SC (1996). Microbially influenced corrosion of Fe, Ni, Cu, Al, and Ti based weldments in a marine environment. *CORROSION/1996*, Paper No. 274. Houston, TX: NACE International.

Dowling NJE, Franklin M, White DC, Lee CH, Lundin C (1989). The effect of microbiologically influenced corrosion on stainless steel weldments in artificial seawater. *CORROSION/1989*, Paper No. 187. Houston, TX: NACE International.

Dowling NJE, White DC, Buchanan RA, Danko JC, Vass A, Brooks S, Gary L (1990). Microbiologically influenced corrosion of 6% molybdenum stainless steels and AISI 316: comparison with ferric chloride testing. *CORROSION/1990*, Paper No. 532. Houston, TX: NACE International.

Enos DG, Taylor SR (1996). Influence of sulfate-reducing bacteria on alloy 625 and austenitic stainless steel weldments. *Corros. Sci.*, **52**(11): 831–842.

Felder CM, Stein AA (1994). Microbiologically influenced corrosion of stainless steel weld and base metal – four year test results. *CORROSION/1994*, Paper No. 275. Houston, TX: NACE International.

Gouda V, Banat I, Riad W, Mansour S (1993). Microbiologically induced corrosion of UNS N04400 in seawater. *Corrosion*, **49**(1): 63.

Hanford R (1990). Nuclear plant remedies for MIC with 'super' stainless steel. *CORROSION/ 1990*, Paper No. 535. Houston, TX: NACE International.

Horn J, Martin S, Masterson B (2001). Evidence of biogenic corrosion of titanium after exposure to a continuous culture of *Thiobacillus ferrooxidans* grown in thiosulfate medium. *CORROSION/2001*, Paper No. 01259. Houston, TX: NACE International.

Johnsen R, Bardal E (1985). Cathodic protection of different stainless steels in natural seawater. *Corrosion*, **41**(5): 296.

Kearns J, Borenstein S (1991). Microbially influenced corrosion testing of welded stainless alloys for nuclear power plant service water systems. *CORROSION/1991*, Paper No. 279. Houston, TX: NACE International.

Kobrin G (1976). Corrosion by microbiological organisms in natural waters. *Mater. Perform.*, **15**(7): 38–42.

Kovach CW, Redmond JD (1997). High performance stainless steels and microbiologically influenced corrosion. *Avesta Sheffield Corrosion Management*, Paper No. 1, p. 9.

Licina GJ, Anderson S, Maner MK (1991). In-plant electrochemical studies of service water system materials. *CORROSION/1991*, Paper No: 277. Houston, TX: NACE International.

Little BJ, Wagner PA, Jacobus J (1988). The impact of sulfate-reducing bacteria on welded copper-nickel seawater piping systems. *CORROSION/1988*, Paper No. 81. Houston, TX: NACE International.

Little BJ, Wagner PA, Mansfeld F (1991). Microbiologically influenced corrosion of metals and alloys. *Int. Mater. Rev.*, **36**(6): 253–272.

Little B, Wagner P, Ray R (1992). An experimental evaluation of titanium's resistance to microbiologically influenced corrosion. *CORROSION/1992*, Paper No. 173. Houston, TX: NACE International.

Little B, Wagner PA, Ray RI (1993). An evaluation of titanium exposed to thermophillic and marine biofilms. *CORROSION/1993*, Paper No. 308. Houston, TX: NACE International.

Little B, Wagner P, Ray R, McNeil M (1990). Microbiologically influenced corrosion in copper and nickel piping systems. *Mar. Technol. Soc. J.*, **24**(3): 10.

Mansfeld F, Little BJ (1992). Microbiologically influenced corrosion of copper-based materials exposed to natural seawater. *Electrochim. Acta*, **37**(12): 2291–2297.

Martin S, Horn J, Carrillo C, (2004). Micron-scale MIC of alloy 22 after long term incubation in saturated nuclear waste repository microcosms. *CORROSION/2004*, Paper No. 04596. Houston, TX: NACE International.

McNeil MB, Jones JM, Little BJ (1991). Production of sulfide minerals by sulfate-reducing bacteria during microbiologically influenced corrosion of copper. *Corrosion*, **47**(9): 674–677.

McNeil MB, Odom AL (1992). Prediction of sulfiding corrosion of alloys induced by consortia containing sulfate reducing bacteria (SRB). *International Symposium on Microbiologically Influenced Corrosion (MIC) Testing*, Miami, FL: ASTM.

Davis JR (ed) (1998). *Metals Handbook*, 2nd edn. Materials Park, OH: ASM International.

Nandakumar K, Sreekumari KR, Kikuchi Y (2002). Antibacterial properties of magnesium alloy AZ31B: *in-vitro* studies using the biofilm-forming bacterium *Pseudomonas* sp. *Biofouling*, **18**(2): 129–135.

Olefjord I, Wegrelius L (1993). Passivation of high alloyed stainless steels. *Symposium on Modifications of Passive Films*, Paris, France.

Paradies HH, Hansel I, Fischer W, Wagner D (1990). *The Occurrence of a Severe Failure of Water Supplies in a County Hospital*. International Copper Research Institute, Project No. 404.

Pope DH, Duquette DJ, Johannes AH, Wayner PC (1984). Microbiologically influenced corrosion of industrial alloys. *Mater. Perform.*, **23**(4): 14–18.

Pope DH (1987). *Microbial Corrosion in Fossil-Fired Power Plants – A Study of Microbiologically Influenced Corrosion and a Practical Guide for its Treatment and Prevention*. Palo Alto, CA: Electric Power Research Institute.

Salvarezza RC, Videla HA (1984). Microbiological corrosion in fuel storage tanks. Part I – anodic behavior. *Acta Cient. Venez.*, **35**: 244–247.

Schumacher M (1979). *Seawater Corrosion Handbook*. Park Ridge, NJ: Noyes Data Corporation.

Schutz RW (1991). A case for titanium's resistance to microbiologically influenced corrosion. *Mater. Perform.*, **30**(1): 58–61.

Scott PJB, Davies M (1989). Microbiologically influenced corrosion of alloy 904L. *Mater. Perform.*, **28**(5): 57–63.

Shreir LL, Jarman RA, Burstein GT (eds) (1994). *Corrosion – Metal/Environment Reactions*, 3rd edn. London: Butterworth-Heinemann, Vol. 1.

Stein AA (1991). Metallurgical factors affecting the resistance of 300 series stainless steels to microbiologically influenced corrosion. *CORROSION/1991*, Paper No. 107. Houston, TX: NACE International.

Sreekumari KR, Nandakumar K, Kikuchi Y (2004). Effect of metal microstructure on bacterial attachment: a contributing factor for preferential MIC attack of welds. *CORROSION/2004*, Paper No: 04597. Houston, TX: NACE International.

Sreekumari KR, Nandakumar K, Yokota T, Kikuchi Y (2002). Laboratory assay of antibacterial properties and corrosion behavior of silver alloyed AISI type 304 stainless steel and its welds. *CORROSION/2002*, Paper No. 02469. Houston, TX: NACE International.

Uhlig HH, Revie WR (1985). *Corrosion and Corrosion Control: An Introduction to Corrosion Science and Engineering*, 3rd edn. New York, NY: Wiley-Interscience.

Videla HA, Gómez de Saravia SG, de Mel MFL, Guiamet PS (1990). Bioelectrochemical assessment of biofilm effects on MIC of two different steels of industrial interest. *CORROSION/1990*, Paper No. 123. Houston, TX: NACE International.

Wagner P, Little B, Ray R, Jones-Meehan J (1992). Investigations of microbiologically influenced corrosion using environmental scanning electron microscopy. *CORROSION/1992*, Paper No. 185. Houston, TX: NACE International.

Walsh DW (1999). The effects of microstructure on MIC susceptibility in high strength aluminum alloys. *CORROSION/1999*, Paper No. 187. Houston, TX: NACE International.

Walsh DW, Seagoe J, Williams L (1992). Microbiologically influenced corrosion of stainless steel weldments: attachment and film evolution. *CORROSION/1992*, Paper No. 165. Houston, TX: NACE International.

Walsh DW, Willis ER, VanDiepen T (1995a). Susceptibility of low alloy steel weldments to microbiologically influenced corrosion: attachment and pitting initiation. *International Conference on Microbially Influenced Corrosion*, **61**: 18.

Walsh DW, Willis ER, Van Diepen T (1995b). The effects of microstructural changes caused by welding on microbiologically influenced corrosion: material and process implications. *CORROSION/1995*, Paper No. 221. Houston, TX: NACE International.

Ward GL, Anderson S, Maner MK (1991). MIC and corrosion of service water system materials. *CORROSION/1991*, Paper No. 278. Houston, TX: NACE International.

Willis ER, Walsh DS (1995). The effect of inclusion composition and morphology on microbiologically influenced corrosion in low alloy steels. *CORROSION/1995*, Paper No. 220. Houston, TX: NACE International.

Yaniv AE, Lumsden JB, Staehle RW (1977). The composition of passive films on ferritic stainless steels. *J. Electrochem. Soc.*, **124**(4): 490–496.

Zhang HJ, Dexter SC (1995). Effect of biofilms on crevice corrosion of stainless steels in coastal seawater. *Corrosion*, **1**(1): 56–66.

Chapter 7

Design Features that Determine Microbiologically Influenced Corrosion

INTRODUCTION

In general, design practices that make materials more vulnerable to other types of corrosion also encourage MIC, for example, surface contamination, butt and socket welds, gasketed joints, lap joints and crevices, flange torquing, pipe threading, and residual stresses. In addition, environments and operating conditions that encourage microbial growth are conducive to MIC. Stagnant and low-flow (3 ft s^{-1}) conditions provide optimum environments for the growth of microorganisms and MIC.

HYDROTEST PROCEDURES

According to Stein (1993), "A system's first encounter with MIC usually occurs during its initial exposure to an aqueous environment, such as during hydrotest, wet lay-up, or moist soil." In contrast, Lutey (2001) maintains that, "There is limited factual documentation available to substantiate the claim that MIC often is established during preoperational phases of plant construction or during the preconditioning and hydrostatic testing phases when water is first introduced into the system." Despite differing opinions on the significance of inappropriate hydrotest procedures to the occurrence of MIC, both authors stress the need for special attention to the selection of hydrotest waters and procedures to prevent MIC. As a result of widely publicized failures due to poor hydrotest waters and practices,

Microbiologically Influenced Corrosion By Brenda J. Little and Jason S. Lee
Published 2007 by John Wiley & Sons, Inc.

Kobrin (1976) developed a list of prioritized, preferred hydrotest waters and practices for chemical process industries:

- First choice — Use demineralized water, drain, and dry as soon as possible after hydrotesting. (MIC has been reported in systems maintained with stagnant demineralized water.)
- Second choice — Use high-purity steam condensate with early draining and drying.
- Last choice — Use natural freshwater (river, pond, canal, well, etc.), drain immediately, flush with demineralized water or steam condensate, blow or mop-dry within 3–5 days.

Drying can be accomplished using compressed air, or preferably dry nitrogen gas. Hydrostatic testing is usually performed in progressive stages, sequentially testing each segment of a system. Segments should be drained within days of hydrotesting. Chlorinated potable waters, estuarine or brackish water, and natural seawater have been used as hydrotest waters for 300 series austenitic stainless steels (UNS S30400, S30403, S31600, S31603) with disastrous results (Kobrin, 1976; Borenstein and Lindsay, 1993; Stoecker, 1981). Not only do these waters contain microorganisms, but they also contain chloride. In 1995, the American Petroleum Institute issued a paragraph dealing with hydrotest procedures for 300 series stainless steels that included draining, drying, and the possibility of adding corrosion inhibitors to reduce the risk of MIC. The use of traditional corrosion inhibitors such as chromates, phosphates, molybdates, zinc, and azoles has not been effective in preventing MIC (Lutey, 2001). Prasad (2003) described special blends of corrosion inhibitors in combination with biocides that were effective in protecting carbon steel (undesignated) corrosion coupons from MIC when using Galveston Bay seawater as hydrotest water.

Stagnant storm water, rainwater, or melted snow in pipes or open tanks provide environments for microbial growth. Many operating systems require stagnant water that cannot be drained or dried. For example, some fire protection systems remain charged with water for long periods of time between uses.

In addition to time of stagnation, the quality of makeup water is also a primary consideration that determines the potential for MIC. Important water-quality criteria identified by Lutey (2001) include low organic content, minimum soluble iron, minimum microflora and fauna, minimum turbidity and suspended solids, minimum dissolved gases, including H_2S, NH_3, O_2, and CO_2, and essentially no hydrocarbons such as oils and greases. Makeup waters are usually surface sources such as lakes, reservoirs, rivers, and groundwaters that are treated by some combination of filtration, pH adjustment, sulfate ion removal, iron sequestration, ion exchange, aeration, deaeration, nitrate/phosphate removal, softening, coagulation/flocculation/clarification, and oil–water separation.

FLOW

Cell coverage of the substratum is strongly influenced by flow with higher shear stresses resulting in lower absorbed cell densities (Characklis, 1990). Licina (1988) suggested that a flow rate of 1 m s^{-1} (3 ft s^{-1}) should be maintained to discourage thick biofilm formation. Continuous flow is preferable to intermittent flow. Dead leg and bypass circuits should be avoided wherever possible. When they are required and when standby components are part of the system, the system design should include drains and the capability for periodic draining and cleaning. Side-stream filtration and in-line filters should be included in the design when makeup water contains high levels of suspended solids. Provisions should be made to purge accumulations of suspended solids from the system when it is not possible to prevent their accumulation by insertion of pigs, air bumping, sand-jetting, or high-pressure water spray.

SUMMARY

Prevention of MIC begins at the design stage. The system design should include provisions for accessibility for draining, cleaning, and water treatment. Components that typically remain stagnant for long periods of time should be designed to facilitate cleaning or flushing. Where possible, the system design should provide control of the flow velocity that will limit bacterial growth. Engineering designs that incorporate drains, eliminate traps for stagnant water, reduce the number of bends and elbows, and specify gasket materials that do not wick reduce the potential for MIC.

REFERENCES

Borenstein SW, Lindsay PB (1993). MIC failure analyses. In: Kobrin G (ed) *A Practical Manual on Microbiologically Influenced Corrosion*. Houston, TX: NACE International, pp. 21–24.

Characklis WC (1990). Biofilm process. In: Characklis WG, Marshall KC (eds) *Biofilms*. New York: John Wiley, p. 217.

Kobrin G (1976). Corrosion by microbiological organisms in natural waters. *Mater. Perform.*, **15**(7): 38–43.

Licina GT (1988). *Sourcebook for Microbiologically Influenced Corrosion*. Electric Power Institute, ERRI NP-6815 – D, Palo Alto, CA, pp. 3–5.

Lutey RW (2001). Treatment for the mitigation of MIC. In: Stoecker J (ed) *A Practical Manual on Microbiologically Influenced Corrosion*. Houston, TX: NACE International, pp. 9.1–9.30.

Prasad R, (2003). Chemical treatment options for hydrotest water to control corrosion and bacterial growth. CORROSION/2003, Paper No. 03572, Houston, TX: NACE International.

Stein AA (1993). MIC in the power industry. In: Kobrin G (ed) *A Practical Manual on Microbiologically Influenced Corrosion*. Houston, TX: NACE International, pp. 21–24.

Stoecker JG (1981). Penetration of stainless steel following hydrostatic test. *Mater. Perform.*, **20**(8): 43–44.

Chapter 8

Case Histories

INTRODUCTION

Original case histories of microbiologically influenced corrosion (MIC) have been published in peer-reviewed journals, symposia proceedings, and edited collections (Dexter, 1986; Kobrin, 1993; Stoecker, 2001; Little, 2002). Case histories have been selected for inclusion in the following sections because they typify frequently occurring failures or because of their health/economic/safety relevance. The challenge in attempting to present an overview of MIC case histories is one of organization. In the past, case histories have been organized by metal/alloy, specific industry problems, or by failure mode, for example, materials vulnerable to sulfate-reducing bacteria (SRB). In the following sections, the case histories are arranged generally by environments, for example, subterranean, atmospheric, and marine exposure. Additional case histories are presented in separate categories for more unique problems, for example, nuclear waste storage that can include both freshwater immersion and subterranean burial.

GENERIC ENVIRONMENTS

Subterranean

The first recorded case of MIC was failure of cast iron water pipes buried in wet, anaerobic soil (von Wolzogen Kuhr and van der Vlugt, 1934). Since that observation, many investigators have reported the effects of bacteria, especially SRB, acid-producing bacteria (APB), and fungi on MIC in subterranean environments (Li et al., 2003; Pope et al., 1991).

External Pipeline Surfaces

Fuel and hazardous materials   In the United States, the pipeline infrastructure includes roughly 2.2 million miles of pipe, including 156,000 miles of hazardous

Microbiologically Influenced Corrosion By Brenda J. Little and Jason S. Lee
Published 2007 by John Wiley & Sons, Inc.

liquid transmission pipelines, 325,000 miles of natural-gas transmission pipelines, and 1.7 million miles of natural-gas distribution pipelines. Gathering pipelines collect products from sources, such as wells on land or offshore or tankers, and move the product to storage or processing areas.

Most gathering, transmission, and distribution pipelines are carbon steel and the external surfaces are protected by coatings (Table 8-1) (Sloan, 2001), cathodic protection, or a combination of both. Pope and Morris (1995) indicated that almost all cases of MIC on buried external surfaces are associated with disbonded coatings or other areas shielded from cathodic protection and most MIC failures occur in wet clays. Jack (2001) reported that NOVA Corporation of Alberta, a major transporter of natural gas, had experienced significant external corrosion on a specific section of undesignated pipeline running through remote muskeg regions of Alberta. Despite apparent cathodic protection levels in excess of -1000 mV (Cu/CuSO$_4$), corrosion rates of 0.7 mm year^{-1} were reported under disbonded polyolefin tape coatings in regions of wet clay. Corrosion products at the site were rich in iron sulfides and iron (II) carbonates. Sulfate-reducing bacteria were found in the corrosion products and trapped water under disbonded coatings. Jack (1991) demonstrated that the tape adhesive/primer and the disbonded coating were sources of nutrients for the SRB.

Water   Alanis et al. (1986) investigated the causes of corrosion in underground copper-zinc brass (undesignated) pipes (91.6 percent copper, 8.22 percent zinc) for a drinking water distribution system. The pipes, less than 2 years in service at the time of perforation, were buried in wet clay soil in a subtropical climate. They concluded that the failures were due to a combination of SRB and aggressive anions in the soil.

In the 1960s, the UK water industry changed from using spun gray iron pipes to using spun ductile iron pipes (Table 8-2) (King et al., 1986). Ductile iron has superior strength and ductility and service experience allowed manufacturers to reduce wall thickness to 70 percent of equivalent gray iron. However, the corrosion performance of this new, mechanically superior material was not as good as expected. King et al. (1986) conducted a 2-year comparative study of the two undesignated materials and concluded that both ductile and gray iron (undesignated) suffered extensive corrosion by SRB. Despite the differences in metallurgical structures, chemistry, and mechanical properties, the materials corroded at similar rates—1250 to 1500 μm year^{-1} (King et al., 1986). In laboratory experiments, Kasahara and Kajiyama (1986) confirmed the role of SRB in the corrosion of undesignated ductile iron exposed to anaerobic soils.

Chung et al. (2001) reported MIC in buried 304L stainless-steel (UNS S30403) piping with 0.22-in. thick, tape-wrapped 316L stainless-steel (UNS S31603) field welds buried underground. Large open cavities developed in the outside pipe surface, primarily in the base metal along the weld crown, in less than 9 months as a result of microorganisms in water trapped under the tape coating.

Electric Cables

Pintado and Montero (1992) reported failures of underground medium-tension electric cables due to bacteria and fungi. The typical structure of a medium-tension

TABLE 8-1   Types of Pipeline Coatings

Pipe coating	Desirable characteristics	Limitations
Coal tar enamels	80+ years of use	Limited manufacturers
	Minimum holiday susceptibility	Limited applicators
	Low current requirements	Health and air quality concerns
	Good resistance to cathodic disbondment	Change in allowable reinforcements
	Good adhesion to steel	
Mill-applied tape systems	30+ years of use	Handling restrictions —shipping and installation
	Minimum holiday susceptibility	UV and thermal blistering—storage potential
	Ease of application	Shielding CP from soil
	Good adhesion to steel	Stress disbondment
	Low energy required for application	
Crosshead-extruded polyolefin with asphalt/ butyl adhesive	40+ years of use	Minimum adhesion to steel
	Minimum holiday susceptibility	Limited storage (except with carbon black)
	Low current requirements	Tendency for tear to propagate along pipe length
	Ease of application	
	Nonpolluting	
	Low energy required for application	
Dual-side-extruded polyolefin with butyl adhesive	25 years of use	Difficult to remove coating
	Minimum holiday susceptibility	Limited applicators
	Low current requirements	
	Excellent resistance to cathodic disbondment	
	Good adhesion to steel	
	Ease of application	
	Nonpolluting	
	Low energy required for application	

TABLE 8-1   *Continued*

Pipe coating	Desirable characteristics	Limitations
Fusion-bonded	35+ years of use	Exacting application parameters
	Low current requirements	High application temperature
	Excellent resistance to cathodic disbondment	Subject to steel pipe surface imperfections
	Excellent adhesion to steel	Lower impact and abrasion resistance
	Excellent resistance to hydrocarbons	High moisture absorption
Multilayer epoxy/ extruded polyolefin systems	Lowest current requirements	Limited applicators
	Highest resistance to cathodic disbondment	Exacting application parameters
	Excellent adhesion to steel	Higher initial cost
	Excellent resistance to hydrocarbons	Possible shielding of CP current
	High impact and abrasion resistance	

Source: Sloan (2001, © NACE International).

TABLE 8-2   Composite of Grey and Ductile Irons (wt.%)

Minor elements	Mean	Max	Min
Grey iron			
Ni	00.08	0.14	0.06
Cr	0.08	0.10	0.05
Cu	0.21	0.26	0.10
Mg	0.04	0.05	0.03
Mo	0.02	0.02	0.01
Ca/Ti	<0.015	–	–
Ce	ND		

Source: King (1986. © NACE International).

electric cable (Figure 8-1) consists of a protective outer layer of bitumen-impregnated jute fiber. A second layer of steel provides mechanical strength and several layers of bituminized jute located between the three phases of the cable. Each conductor is wrapped with insulating paper or cloth embedded in bitumen, a lead sheath, and layers of oil-impregnated insulating paper. These layers are in direct contact with

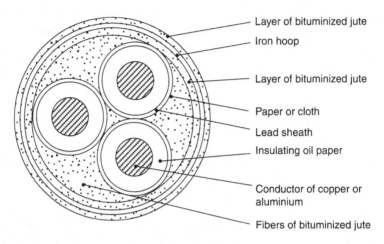

Layer of bituminized jute
Iron hoop
Layer of bituminized jute
Paper or cloth
Lead sheath
Insulating oil paper
Conductor of copper or aluminium
Fibers of bituminized jute

FIGURE 8-1   Scheme of medium-tension electric cable structure. (Reprinted from Pintado and Montero, 1992, with permission from Elsevier.)

copper or aluminum cables. The authors determined that the lead sheaths had been attacked by several species of *Bacillus* and fungi (*Penicillium, Trichoderma,* and *Aspergillus*). They demonstrated that the bacteria and fungi were contaminants of the cable and not from the underground environment. They further demonstrated that the cellulose fibers and bitumen used in the cable construction could support the growth of the isolates, and the degradation of the structural materials caused failure of the cables.

Atmospheric

The availability of water and nutrients influences the distribution and growth of microorganisms. Fungi are the most desiccant-resistant microorganisms and can remain active at $a_w = 0.6$ whereas few bacteria remain active below $a_w = 0.9$. For this reason, most atmospheric MIC is due to the activities of fungi. Fungi are ubiquitous in atmospheric, aquatic, and soil environments where they assimilate organic materials and produce organic acids. Atmospheric biodeterioration due to fungi has been documented for the following: cellulosic materials (paper, composition board, and wood): communication wire; cable splices; telephone cable; cable sheaths; photographic film; polyvinyl chloride films; sonar diaphragm coatings; map coatings; paints; metals; greases; waxes; lubricants; adhesives; asphalt; hydraulic fluids; rain repellents; textiles (cotton and wool); vinyl jackets; leather shoes; feathers and down; natural and synthetic rubber; optical instruments; mechanical, electronic, and electric equipment (radar, radio, flight instruments, wire strain gages, and helicopter rotors); hammocks; tape; thermal insulation; brick masonry and concrete; medicines; and museum valuables.

Ship Holds

Stranger-Johannessen (1984) reported that UNS G10200 carbon steel ship holds carrying cereals, woods, and other dry cargoes were severely corroded within months. Heavy pitting and reduced thickness of the steel plate were observed. A microbiological investigation showed that the corrosion products were populated with viable fungi. Originally, the holds had been coated with chlorinated rubber paint. Laboratory tests demonstrated that the paint provided nutrients for fungal growth and corrosion resulted from acidic metabolic by-products.

Aircraft

Numerous reports document fungal growth in passenger compartments of in-service aircraft coated with polyurethane paint (Figure 8-2 *a*, *b*). Little et al. (2000) documented the following eight fungal genera associated with H-53 aircraft: *Pestolotia, Trichoderma, Epicoccum, Phoma, Stemphylium, Hormodendrum* (also known as *Cladosporium*), *Penicillium,* and *Aureobasidium.* Organisms were cultured from virtually all interior surfaces, including primer- and polyurethane-coated 2024 T-6 aluminum (UNS A92024), fiberglass structural members, caulking, synthetic fabrics, wiring, and air-conditioning ducts. Distribution of organisms was not limited to standing water/fluids. Concentration of organisms did depend on the availability of nutrients and water. The authors determined that the fungi did not degrade the polyurethane topcoat, but the organisms derived nutrients from hydraulic fluid, deposited on surfaces during the normal operation of the helicopter, and a corrosion-inhibiting compound applied to surfaces during maintenance. The topcoat from a piece of peeling paint was uniformly colonized by fungi (Figure 8-3 *a–e*). Environmental scanning electron microscopy (ESEM) was used to demonstrate that fungi penetrated the polyurethane topcoat and caused disbonding between topcoat and primer. There were no indications that the organisms penetrated the chromate conversion coating. Trick and Keil (1999) indicated that the chromate acted as a biocide.

Little et al. (2000) found that glossy finish polyurethane fouled more readily than the same formulation with a flat finish. Aged paint fouled more rapidly than new coatings. Laboratory tests demonstrated that in the presence of hydraulic fluid or corrosion-inhibiting compounds, most of the isolates caused localized corrosion of bare A92024 T-6 aluminum (Figure 8-4 *a*, *b*).

Stranger-Johannessen and Norgaard (1991) demonstrated that microorganisms do not need to be attached to grow on or penetrate a coating to cause blistering and loss of adhesion. Instead, microorganisms could act on the coating from outside by excreting metabolites that can dissolve or react with constituents in the coating. The properties of the coating may be changed in such a way that it becomes more prone to blistering. They found that coatings containing zinc chromate killed fungi and no growth was observed on chromate-coated surfaces.

In a joint project with eight European aircraft manufacturers, Hagenauer et al. (1994) investigated the involvement of microorganisms in corrosion damage in

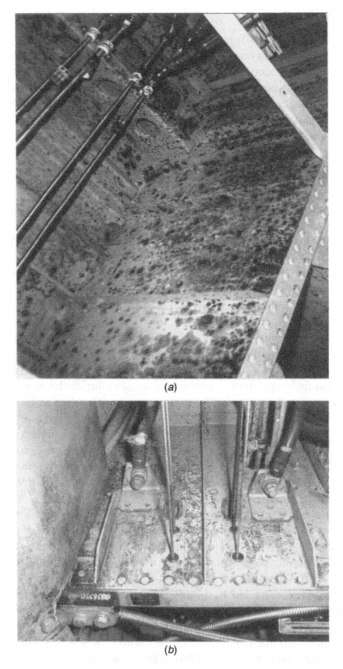

(a)

(b)

FIGURE 8-2a, b Fungal growth in passenger compartments of in-service aircraft coated with polyurethane paint. (Little et al.. 2000.)

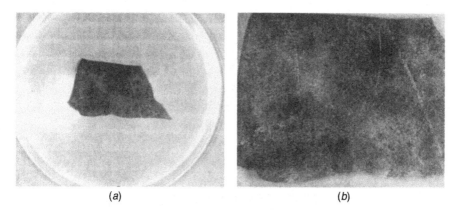

(a)   (b)

FIGURE 8-3*a*, *b*   Topside of paint topcoat peeling uniformly colonized by fungi at (*a*) 1× and (*b*) 3× magnification. (Little et al., 2000.)

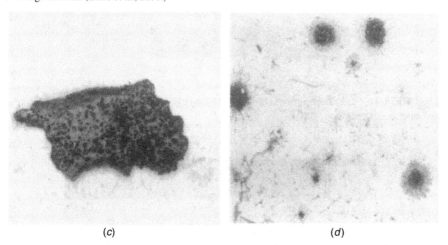

(c)   (d)

FIGURE 8-3*c*, *d*   Underside of paint topcoat peeling colonized by fungi at (*c*) 2× and (*d*) 40× magnification. (Little et al., 2000.)

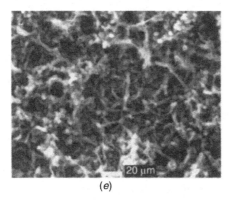

(e)

FIGURE 8-3*e*   ESEM micrograph of fungal mass on paint topcoat. (Little et al., 2000.)

(a)

FIGURE 8-4 a   Localized corrosion of bare 2024 T-6 (UNS A92024) aluminum in the presence of hydraulic fluid. (Little et al., 2000.)

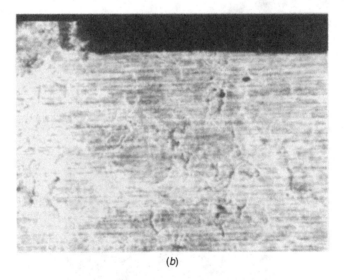

(b)

FIGURE 8-4b   Zoom in of localized corrosion on bare 2024 T-6 (UNS A92024) aluminum in the presence of hydraulic fluid. (Little et al., 2000.)

aircraft. Forty-six samples were taken from corroded sites on different airplanes, a total of 208 microorganisms were isolated—158 bacteria, 36 yeasts, and 14 mycelium-forming fungi (Table 8-3). Using laboratory testing, the authors were able to conclude that some of the isolates could cause corrosion of aluminum 7075 (UNS A97075). However, the authors clearly state they could not conclude the extent to which the organisms promoted corrosion during operation of the aircraft.

Wire Rope

Little et al. (1995) reported fungus-influenced corrosion of undesignated carbon steel wire rope used in military applications. Laboratory data and field statistics implicated the practice of wrapping in plastic, stored carbon steel highlines, under humid conditions, with MIC. The 1-in. carbon steel wire ropes were composed of six individual strands around an independent wire core coated with a thick maintenance grease and threaded onto wooden spools that were wrapped in brown paper and black plastic prior to storage. Wire rope may be stored for weeks to months before being used. Fungal growth was observed on interiors of some wooden spools stored outdoors and corrosion was most severe on wraps of wire in direct contact with the wooden spool flanges (Figure 8-5a, b). *Aspergillus niger* and *Penicillium* sp. were isolated from wooden spool flanges. Isolates could not be maintained on the protective grease, that is, these fungi could not break down the maintenance coating to obtain nutrients. Instead, fungi growing on wood produced copious amounts of CO_2, and the pH of any associated condensate was acidic enough to dissolve the maintenance grease and produce localized corrosion as demonstrated in the laboratory (Figure 8-6a–c).

Building Materials

Microbiologically influenced corrosion is responsible for the deterioration of concrete, sandstone, bricks, terra-cotta, plaster, mortar, and marble exposed under

TABLE 8-3 Results of Sampling in Seven Different Aeroplanes

Aircraft	Number of samples	Number of isolates	Sampling area
1. Airbus A 300	3	15	Toilet, pass door, and mid-cargo door areas
2. Boeing 727	1	9	Pass door sill
3. Boeing 727	2	28	Forward cargo compartment
4. Helicopter	3	11	Door to fuselage, cabin floor, and tail boom
5. Tornado	2	7	Left wing
6. Tornado	15	23	Wings, fuel tank door, and fuel tank area
7. Airbus A 300	20	115	Forward cargo, toilet area, cargo area, and cabin floor
Total	46	208	

Source: Reprinted from Hagenauer et al. (1994), with permission from Wiley-VCH Verlag GmbH & Co KG.

 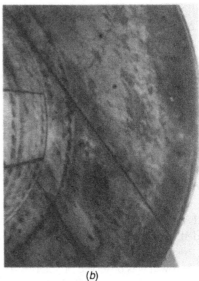

(a)   (b)

FIGURE 8-5a, b   Fungal growth observed on interiors of wooden spools stored outdoors. (Little et al., 1995.)

atmospheric conditions. Microbial colonization and biodeterioration of building materials are linked to environmental conditions (Ortega-Calvo et al., 2001). The most significant parameters are moisture, temperature, light, and the chemical nature of the substratum. The initial colonizers require only water and a minimal supply of mineral salts. Saiz-Jimenez (2001) concluded that the duration of wetness is more important than the frequency of wetting for predisposing a surface to colonization. Colonization is more rapid in the presence of overhanging trees, bird droppings, agricultural fertilizers, and pollution-derived nitrogen oxides. Rough or porous surfaces facilitate attachment of airborne contaminants and accumulation of nutrients. Most stones collect organic material that can maintain large populations of bacteria. Many of the most frequently used building materials, limestone, marble, sandstone, and mortar, are carbonates and are unstable in acid solutions. Therefore, any organisms that produce organic or inorganic acids can cause the dissolution of carbonates. The most conspicuous attacks are by phototrophic organisms, meaning organisms that obtain their energy for growth from light, and include cyanobacteria, algae, and lichens.

Fungi can thrive on building stones and their growth can induce mechanical or chemical changes. Decorative tumuli have been located in Japan. Decorative tumuli are burial mounds with mural paintings. When tumuli were excavated, there were no signs of microbial growth; however, once excavation took place, surface stains, discolorations, and flaking developed in the relative humidity of typically 95 to 100 percent. After 5 days, the numbers of fungi increased by an order of magnitude.

Videla et al. (2003) investigated the effects of rock decay on the Church of Vera Cruz in Medellin, Columbia. Church construction was started in 1682 and its front is built of peridotite, a rock containing iron and magnesium minerals such as olivine and pyroxene. The authors surmised that microbial activities synergistically enhanced atmospheric effects on rock decay. Gaylarde et al. (2001) investigated buildings at Uxmal and Tulum, two Mayan sites in the Yucatan peninsula. The buildings are composed of limestone. Gray-black discolorations on exposed walls and copious green biofilms on inner walls were observed. They isolated cyanobacteria, heterotrophic bacteria, and filamentous fungi. Mechanisms for the surface discolorations included acid attack by bacteria, fungi, and algae and direct boring by fungi.

Diercks et al. (1991) documented the degradation of stone buildings in Germany due to nitrifying organisms that oxidize ammonia and nitrate to produce nitric acid. Nitric acid dissolved carbonaceous binding materials and modified clay-like components. Nitrates formed by neutralizing nitric acid were hygroscopic and retained moisture in the stone even at low relative humidity.

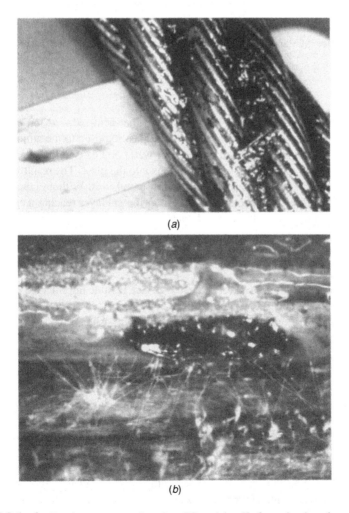

(a)

(b)

FIGURE 8-6a, b   Fungi grown on wood produce CO_2 and the pH of associated condensate is acidic enough to dissolve maintenance grease and produce localized corrosion. (Little et al., 1995.)

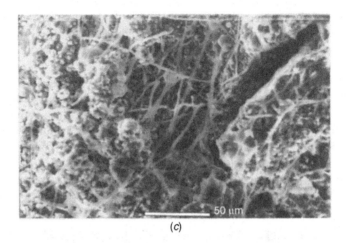
(c)

FIGURE 8-6c   SEM image of fungi grown on wood and the associated localized corrosion. (Little et al., 1995.)

Glass

Drewello and Weissmann (1997) demonstrated that acid-induced ion exchange by fungal mycelia was important for corrosion of glass exposed to the atmosphere. The corrosive attack depended on the presence of water and caused the release of metal ions as well as the diffusion of hydronium ions into the glass. The result was an inner layer of hydrated silica gel largely depleted in metal ions. Without cracks the layer was protective. However, cracks can occur after the gel layer reaches a critical thickness, causing a loss of both glass and coating.

Marine

The materials most often used for seawater service include unprotected carbon steel, cathodically protected and coated carbon, low alloy steel, stainless steels, copper- and nickel-based alloys, titanium, concrete, and composites.

Iron and Steel

Unprotected steel   Cullimore and Johnson (1999, 2000) identified MIC on the sunken remains of the RMS Titanic. They determined that accelerated corrosion was associated with crystalline structures composed of microbial communities connected with water channels. They isolated SRB, iron-related, slime-forming, denitrifying, nitrifying, sulfur-oxidizing, and heterotrophic bacteria as well as species of fungi and algae. In the 1999 survey, it was determined that 65 percent of the side shell and 25 percent of the bulkhead plating was coated with microbiologically produced tubercles.

Sanders and Hamilton (1986) analyzed anaerobic MIC in North Sea oil exploration (Figure 8-7) and defined two distinct forms of SRB-mediated corrosion: (1) pitting

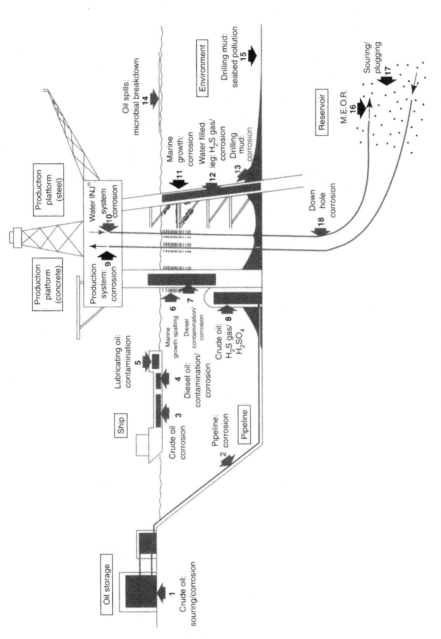

FIGURE 8-7 Situations where microbes are involved in the operations of the North Sea oil industry. (Sanders and Hamilton, 1986. © NACE International.)

163

caused by SRB growing in the biofilm on metal surfaces, and (2) sulfide-induced stress corrosion cracking and hydrogen-induced cracking or blistering caused by hydrogen permeation in high dissolved sulfide concentrations.

Hydrogen embrittlement of carbon steel is a form of corrosion involving the cathodic reaction. Atomic hydrogen generated in the cathodic reaction penetrates the steel, resulting in the loss of ductility. A number of detailed mechanisms have been postulated to account for the embrittlement effect, which, when combined with the presence of local regions of high stress, can result in severe cracking. Hydrogen embrittlement is enhanced by the high levels of hydrogen generated by SRB under certain conditions, as well as that generated by overly aggressive cathodic protection systems. If the levels of cathodic protection are greatly increased to combat SRB corrosion, there is a danger that hydrogen embrittlement may be enhanced. The presence of H_2S, which can be produced by SRB, is known to retard formation of molecular hydrogen on the metal surface and to enhance adsorption of atomic hydrogen by the metal and hydrogen embrittlement.

Accelerated low water corrosion (ALWC) or lowest astronomical tide (LAT) corrosion is an aggressive form of localized corrosion associated with unusually high rates of metal wastage on unprotected, or inadequately protected, steel-sheet pilings. Sheet piles are used as retaining walls, wharfs, and piers. In tidal waters, pilings are exposed to a range of corrosive environments, which can be classified into four zones: splash, tidal, permanent immersion, and bed (Figure 8-8). Corrosion rates in the low water level are typically 0.1 mm year^{-1}. Average corrosion rates in the range 0.3 to 1.2 mm per side per year are typically reported for ALWC (Cheung et al., 1994).

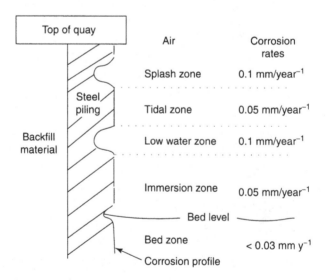

FIGURE 8-8 Typical corrosion profile of steel piling in tidal water. (Reprinted from Cheung et al., 1994, with permission from Elsevier.)

Recognition of ALWC is attributed to Morley and Bruce (1983). However in 1914, Ellis reported corrosion of undesignated steel-joist piles in Hong Kong harbor. Pits and "saucer-shaped depressions" filled with black iron sulfide were found on relatively new structures at and below the low water zone, associated with corrosion rates estimated at 2 mm per side per year. Generation of hydrogen sulfide by "certain marine growths" was cited as the cause. Copenhagen (1954) described a specific low water corrosion problem on undesignated steel-sheet pilings at Cape Town, South Africa. Corrosion rates of 1.25 mm per side per year were observed associated with black, sulfide-rich deposits reported to be "due to the action of sulfate-reducing microorganisms."

The ALWC phenomenon is a global phenomenon, having been reported around the world in all climatic conditions on unprotected steel pilings in contact with saline water (i.e., seawater and brackish water), subject to tidal influences. Accelerated low water corrosion has been reported in the United Kingdom, Scandinavia, Belgium, Germany (Hamburg and harbors located both in the North Sea and the Baltic Sea), Holland, France, the United States, Canada, South Africa, Australia, Cyprus, the United Arab Emigrates, Japan, Ethiopia, and in the southern Caribbean (Granada) (Johnson et al., 1993). A survey of port and harbor authorities in five western European countries conducted in 1994 as part of a study by the European Coal and Steel Communities concluded that at least 13 percent of the ports was affected by ALWC on steel-piled structures (Johnson et al., 1994).

Accelerated low water corrosion has a distinct appearance—patches of lightly adherent, bright orange and black (iron sulfide rich) deposits (Figure 8-9) over a clean, shiny, and pitted steel surface (Beech et al., 1993, 2001, 2004; Cheung et al., 1994). As the pits deepen and become more numerous, they overlap, producing a dishing effect in the metal surface, which ultimately develops into a hole. Corrosion products contain magnetite, iron sulfides, and green rust (an unstable iron oxyhydroxide sulfate complex). Accelerated low water corrosion also produces a specific pattern of damage on steel-sheet piling. In U-shaped piles, the corrosion occurs on the outpans, while in the Z-shaped sheet piles ALWC occurs in the web areas or

FIGURE 8-9   Multilayered surface deposits characteristic of ALWC. (Image provided by I.B. Beech, University of Portsmouth.)

corners (Figure 8-10 a–c). The pattern of damage is similar for particular pile geometries, irrespective of the geographic location of the installation.

The detailed mechanism of ALWC in marine/estuarine environments continues to be a matter of some debate but several research groups have concluded that it is a

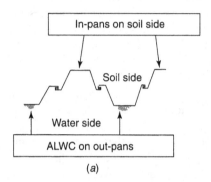

(a)

FIGURE 8-10a   ALWC typically occurs in center of outpans on waterside of U-shaped sheet piles. (Kumar and Stephenson, 2005. © NACE International.)

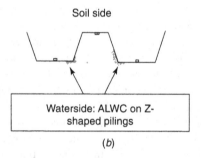

(b)

FIGURE 8-10b   ALWC typically occurs in the web areas or corners on waterside of Z-shaped sheet piles. (Kumar and Stephenson, 2005. © NACE International.)

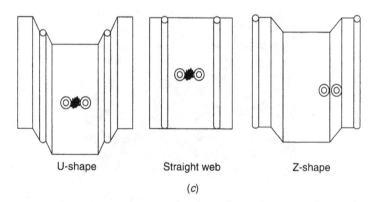

(c)

FIGURE 8-10c   Location of corrosion on U-piles, straight web piles, and Z-piles. (Gubner and Beech, 1999a. © NACE International.)

form of MIC. Gubner and Beech (1999a, b) conducted a statistical assessment of the risk of ALWC in the marine environment using a combination of biological, environmental, chemical, and physical parameters (Table 8-4). They found higher levels of total organic carbon and chloride at ALWC sites (Table 8-5). There were no

TABLE 8-4   Parameters that Can Affect the Corrosivity of Seawater

- Biochemical oxygen demand during 5 days (BOD_5)
- Chemical oxygen demand (COD)
- Total organic carbon (TOC)
- Total aromatic compounds (TAC)
- Polycyclic aromatic hydrocarbons (PAH)
- Inorganic compounds (i.e., PO_4^{2-} NO_3^-, SO_4^{2-}, S_2^-, CN^-, and NH_4^+)
- Heavy metal ions (i.e., Fe, Cr, Cu, Ni, Hg, Zn, and Sn)
- Bacteriological analysis[15]
  - Most probable number (MPN) for SRB
  - MPN for sulfur-oxidizing bacteria (SOB)
  - MPN for chemoorganotrophic bacteria (COT)
  - Total bacterial count

Source: Gubner and Beech (1999a. © NACE International).

TABLE 8-5   Summary of Statistically Significant Parameters Related to the ALWC-Corrosion Phenomenon

Parameter	ALWC S.D.[a] mean value ±	NLWC S.D. mean value ±	One-tailed t-test
Mean tidal range (m)	3.0 ± 2.1	4.9 ± 3.0	$P = 0.047$
Baltic Sea data excluded (m)	3.5 ± 2.1	5.3 ± 2.9	$P = 0.075$
Thickness of corrosion products (mm)	9.7 ± 6.4	6.1 ± 2.8	$P = 0.016$
pH underneath corrosion products	6.2 ± 0.7	6.8 ± 0.7	$P = 0.021$
Redox potential of seawater ($mV_{Ag/AgCl}$)	0 ± 106	54 ± 74	$P = 0.05$
Presence of invertebrates (%)	75%	93%	$P = 0.07$
Presence of algae (%)	90%	67%	$P = 0.06$
Organic carbon (%)	2.2 ± 0.5	1.4 ± 0.5	$P = 0.028$
Organic hydrogen (%)	1.2 ± 0.2	0.8 ± 0.2	$P = 0.041$
Organic nitrogen (%)	0.36 ± 0.06	0.31 ± 0.02	$P = 0.06$
Total organic carbon of seawater (mg dm^{-3} O_2)	14.6 ± 13.5	7.2 ± 7.9	$P = 0.036$
Most probable number of sulfur-oxidizing bacteria (cells g^{-1} dry weight of corrosion products)	$6.5 \times 10^5 \pm$ 2.1×10^6	$3.5 \times 10^3 \pm$ 5.9×10^3	$P = 0.08$

Source: Gubner and Beech (1999a. © NACE International).
[a]Standard deviation.

differences in the suspended matter between the ALWC sites and sites free of the problem. They also eliminated the possibility of cyclic stress as an influence. Corrosion products were softer and easier to remove from the ALWC sites. Biofilms were present in all corrosion products; however, the pH was lower beneath corrosion products at ALWC sites and the numbers of thiobacilli were higher.

Gehrke and Sand (2003) completed a 3-year study of undesignated steel pilings in German marine harbors with and without corrosion. They concluded that the ALWC was due to the combination of SRB and thiobacilli in the fouling layers on the pilings (Figure 8-11). The organisms occurred together, separated by the oxygen gradient in the biofilm. At low tide the biofouling layer was oxygenated whereas at high tide, anaerobic areas developed. The sulfides produced by the SRB in the anaerobic regions and sulfuric acid resulting from the thiobacilli in the aerobic regions combined to produce an extremely corrosive environment.

Kumar and Stephenson (2005) demonstrated that protective coatings, such as coal tar epoxy or glass flake composite, can be used to mitigate ALWC. They

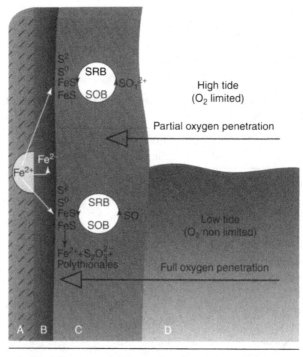

A: Steel pile
B: Black layer
C: Fouling layer
D: Sea water

Upper part: resembles splash zone
Lower part: resembles immersion zone

(Dimensions not proportional)

FIGURE 8-11   Schematic summary on the mechanism of MIC, and ALWC on marine sheet piling structures at the low water level. (Gehrke and Sand, 2003. © NACE International.)

concluded that ALWC can be reduced or eliminated using sacrificial anodes in the immersed zone or impressed current cathodic protection.

Accelerated corrosion has been reported for ASTM A328 sheet pilings in the Duluth-Superior harbor, Minnesota (Marsh et al., 2005). A diver described the corrosion as pockmarks primarily in the 4 ft just below the water surface. The corrosion extended down to about 10 ft, but decreased from 4 to 10 ft. Below 10 ft, there was very little corrosion. Zebra mussel attachment began 10 ft below the surface and extended to the bottom of the sheet pile. An orange coating that tended to cover the pits covered the corroding pockmarks. The corrosion always appeared upstream from the harbor. To date, no mechanism for the accelerated corrosion in this freshwater harbor has been identified.

Cathodically protected alloys   It has been reported that cathodic protection (CP) retards microbial growth because of the alkaline pH generated at the surface. However, numerous investigators have demonstrated a relationship between marine fouling and calcareous deposits on cathodically protected surfaces. For example, Mansfeld et al. (1990) polarized three stainless steels [UNS N08366 (AL6X), S30400, and S31600] and grade 2 titanium (UNS R50400) samples to -850 mV$_{SCE}$ in natural Pacific seawater and followed the formation of surface layers using electrochemical impedance spectroscopy (EIS) and scanning electron microscopy (SEM). The applied potential was removed the day before EIS measurements, so that EIS data were collected at E_{corr} in a frequency range between 65 kHz and 1 or 10 MHz with a signal amplitude of 10 mV. Calcareous deposits and bacteria were evident on the surface after 5 days. Areal coverage of the surfaces could be followed with EIS. For an exposure time of 5 days, EIS data were similar for polarized (Figure 8-12) and unpolarized (Figure 8-13) S30400 samples despite the occurrence of scattered calcareous deposits on the polarized surface. After 13 days there were two time constants in EIS spectra of polarized samples, indicating two surface conditions, that is, covered and uncovered. The occurrence of two time constants corresponding to calcareous deposits and uncovered surface was evident in the frequency dependence of the phase angle (Figure 8-12). After 43 days, the time constants were no longer well-defined due to complete coverage of polarized surfaces with calcareous deposits. In all cases, bacterial cells could be imaged between and on top of calcareous deposits (Figure 8-14*a*, *b*). The size and shape of calcareous deposits on polarized titanium differed from those on polarized stainless steel (Figure 8-15*a*–*c*).

Despite overwhelming evidence that bacteria and calcareous deposits coexist on cathodically protected surfaces, their interrelationships are not understood. Often, microbiological data for cathodically polarized surfaces are confusing and impossible to compare because of differing experimental conditions (laboratory vs. field) and techniques used to evaluate constituents within the biofilm. It is further confusing because Edyvean (1984) reported differences in current density, electrochemistry, calcareous deposits, and biofilm formation due to varying organic content of seawater.

Dexter and Lin (1991) investigated the influence of a preexisting biofilm on the formation of calcareous deposits under cathodic protection in natural seawater. They demonstrated that applied current densities up to 100 μA cm^{-2} did not remove

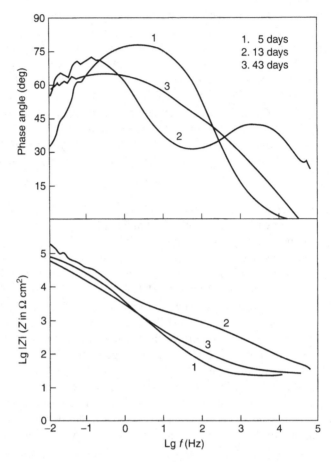

FIGURE 8-12   Impedance spectra for polarized S30400 stainless steel after exposure periods of 5, 13, and 43 days; EIS data taken at E_{corr}. (Mansfeld et al., 1990.)

attached biofilms from C-276 stainless-steel (UNS N10276) surfaces. Both natural Delaware Bay water and a laboratory culture of a marine bacterium changed the morphology of calcareous deposits formed under cathodic polarization at a current density of 100 μA cm^{-2}.

de Mele (1992) reported a dramatic decrease in bacterial populations due to cathodic protection after a 3-h immersion of S30403 stainless steel in pure cultures of *Vibrio alginolyticus* in the laboratory. However, after 9 h there were no significant differences in numbers of SRB on polarized and unpolarized surfaces. Cell numbers were evaluated using epifluorescence microscopy and standard plate counts. Nekoksa and Gutherman (1991) reported that cathodic potentials to -1000 mV$_{SCE}$ caused a decrease in pH and an increase of SRB on undesignated carbon steel and S30403 stainless-steel surfaces. At potentials more negative than -1000 mV$_{SCE}$, the pH became more alkaline and SRB numbers decreased. A study of the influence of SRB in marine sediments

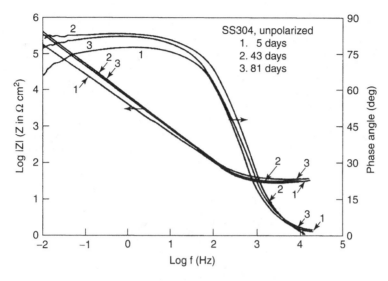

FIGURE 8-13   Impedance spectra for unpolarized S30400 stainless steel. (EIS data taken at E_{corr}) (Mansfeld et al., 1990.)

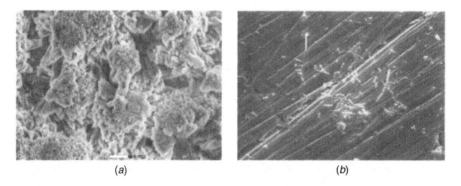

(a)   (b)

FIGURE 8-14 a, b   Bacteria between calcareous deposits on S30400 stainless steel. (Mansfeld et al., 1990.)

using EIS to monitor corrosion and lipid analysis as biological markers, complemented by chemical and microbiological analysis, showed that -880 mV$_{SCE}$ encouraged growth of hydrogenase-positive bacteria in the sediment surrounding the carbon steel and facilitated the growth of other SRB species (Guezennec, 1991).

Because the enumeration technique strongly influences the number of cells one is able to count, and because the number of cells cannot be equated to cellular activity, including sulfate reduction, some investigators have attempted to measure cellular activity directly on cathodically protected surfaces. Maxwell (1986) cathodically protected BS 4360-50D carbon steel (no UNS designation) coupons exposed in the estuarine waters of Aberdeen Harbor using an imposed potential of -950 mV (Cu:CuSO$_4$) and sacrificial anodes. Activities within biofilms were determined

FIGURE 8-15*a–c* (*a, b*) Calcareous deposits on polarized grade 2 titanium (UNS R50400) after 43 days. (*c*) Bacteria between calcareous deposits. (Mansfeld et al., 1990.)

using radiorespirometric methods. Biofilms developed on all cathodically protected and unprotected control substrata. The activities of aerobic and anaerobic bacteria, including SRB, were significantly greater on unprotected coupons. Furthermore, sulfide, a metabolic fingerprint of SRB activity, could only be detected in biofilms on unprotected coupons. These results show that a potential of –950 mV ($Cu:CuSO_4$) does not prevent SRB from developing on cathodically protected surfaces. The lower activity of SRB within biofilms on cathodically protected coupons was not directly caused by any inhibitory effect of the cathodic potential. Instead, the greater activity of SRB on unprotected coupons was the result of production of an extensive corrosion film offering more favorable anaerobic conditions.

Horvath and Novak (1964) studied thermodynamic data with iron in a pH 7 electrolyte saturated with hydrogen sulfide. A potential of –1024 mV_{SCE} was required to achieve cathodic protection. Jack et al. (1986) demonstrated that –1024 mV_{SCE} was capable of providing cathodic protection to carbon steel (undesignated) in the presence of active SRB. The influence of cathodic protection on the growth of SRB and on corrosion of undesignated carbon steel in marine sediments was investigated by Guezennec and Therene (1988). They concluded that a cathodic potential of –880 mV_{SCE} did not appear to be sufficient for protection and that large amounts of cathodically produced hydrogen promoted the growth of SRB in the sediments surrounding the samples. Fischer (1981) conducted laboratory tests in anaerobic, artificial sediments containing

SRB. Results indicated that the criterion of -1024 mV$_{SCE}$ for protection is adequate. Cathodic protection current density was between 4.5 and 12 mA ft^{-2}. Barlo and Berry (1984) concluded that if anaerobic bacterial activity is suspected, a cathodic polarization shift of approximately 200 to 300 mV$_{SCE}$ is required for carbon steel protection.

In summary, bacteria can be demonstrated on cathodically protected surfaces. Cathodic potentials to -1074 mV$_{SCE}$ does not prevent biofilm formation. It has been suggested that actual cell numbers may be related to polarization potential, dissolved organic carbon, or the enumeration technique. Numbers of SRB may be increased or decreased depending on exposure conditions. Carbon steel is considered protected when a potential of -924 mV$_{SCE}$ is achieved. In many cases, the potential is further reduced to -1024 mV$_{SCE}$ to protect the steel from corrosion caused from the activity of SRB. The decreased potential is not applied to prevent the growth of SRB, but is based on a theoretical level that will allow passivity of steel in a sulfide-rich environment produced by SRB. The main consequence of biofilm formation on protected surfaces appears to be an increase in the current density necessary to polarize the metal to the protected potential. The presence of large numbers of cells on cathodically protected surfaces does mean that in the event that cathodic protection is intermittent, discontinuous, or discontinued, the corrosion attack due to the microorganisms will be more aggressive.

Coated carbon steel   Although coatings alone do not prevent MIC, they are used routinely to delay the onset of MIC and other corrosion reactions. Many types of polymeric coatings can be subject to biodegradation. The attack is usually caused by acids or enzymes produced by bacteria or fungi. This often results in selective attack on one or more specific components of a coating system, with consequent increase in porosity and water or other ion transport through the coating, and the formation of blisters, breaches, and disbonded areas. Many of these effects have recently been reviewed by Little et al. (2001). Little et al. (1999) demonstrated that marine bacteria are attracted to corrosion products at coating defects. The microorganisms responsible for damage to coatings may or may not be involved in corrosion initiation under the damaged coating.

Corrosion-resistant and Passive Alloys

At temperatures below 60 °C, resistance to crevice corrosion is the limiting factor for selecting alloys for seawater service and crevice corrosion is the most problematic issue affecting the performance of stainless steels in seawater. Several investigators (Sedricks, 1996; Chandrasekaran and Dexter, 1994; Johnson and Bardal, 1985; Dexter and Gao, 1988; Motoda et al., 1990; Ito et al., 2002) have documented the tendency of biofilms to cause a noble shift, or an ennoblement, in corrosion potential (E_{corr}) of passive alloys exposed in marine environments (Figure 8-16). The alloys tested included, but were not limited to, UNS S30400, S30403, S31600, S31603, S31703, S31725, S31803, N08904, N08367, S44660, S20910, S44735, N10276, platinum, gold, palladium, chromium, titanium, and nickel R50250. Theoretically, potential ennoblement should increase the probability for pitting and crevice corrosion (Zhang and Dexter, 1995) initiation and propagation. However, the authors are unaware of any in-service failures solely attributed to ennoblement in seawater.

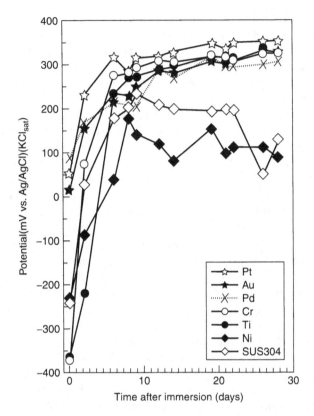

FIGURE 8-16 Potential ennoblement in natural seawater at Futtsu City (Tokyo Bay) from March 15 to April 12 in 2000. (Ito, 2002. © NACE International.)

Copper and Copper–Nickel Alloys

The well-known toxicity of cuprous ions toward living organisms does not mean that the copper-based alloys are immune to MIC. It does mean, however, that only those organisms with a high tolerance for copper are likely to have a substantial effect. *Thiobacillus thiooxidans*, for example, can withstand dissolved copper concentrations as high as 2 percent. Most of the reported cases of microbial corrosion of copper alloys are caused by the production of corrosive substances, for example, CO_2, H_2S, NH_3, and organic or inorganic acids (Little et al., 1991; Pope et al., 1984).

The impact of sulfides on the corrosion of copper alloys has received a considerable amount of attention. Little et al. (1990, 1988) have published reports documenting localized corrosion of 90/10 (UNS C70600) and 80/20 (UNS C71000) copper-nickel alloys by SRB in estuarine environments. Pope (1987, 1986) documented MIC of 90/10 copper-nickel (UNS C70600) and undesignated alloys including admiralty brass (Cu–30Zn—1.5Sn), and aluminum brass (Cu–20Zn–2Al) at electric generating facilities using brackish and fresh cooling waters. Rowlands (1965) reported the failure of aluminum brass (UNS C687700) in polluted seawater containing waterborne sulfides that stimulate pitting and stress corrosion cracking. Copper-nickel alloys C70600 and C17500 and nickel-copper alloy N04400 (detailed later) suffer accelerated corrosion attack in seawater containing 0.01 ppm sulfide after one day's exposure (Gudas and Hack, 1979). In the presence of turbulence, the loosely adherent sulfide film is removed, exposing a fresh copper surface to react with the sulfide ions. For these reasons, turbulence-induced corrosion and the sulfide attack of copper alloys cannot be decoupled easily. In the presence of oxygen, the possible corrosion reactions in a copper sulfide system are extremely complex because of the large number of stable copper sulfides, their differing electrical conductivities and catalytic effects (Ribbe, 1976). Transformations between sulfides, or of sulfides to oxides, result in changes in volume that weaken the attachment scale and oxide subscale leading to spalling.

Little et al. (1988) reported laboratory and field studies of copper-nickel alloys designed to evaluate corrosion in C71000 welds and heat-affected zones in C70600 seawater piping systems exposed to estuarine and seawaters. They showed that welds provide unique environments for the colonization of SRB with the subsequent production of sulfides that affect the weld seam surface, the adjacent flow zone, and the downstream surface of the heat-affected zone. Exposure of sulfide-derived surfaces to fresh, aerated seawater resulted in rapid spalling on the downstream side of weld seams. Bared surfaces became anodic to the sulfide-coated weld root, accelerating corrosion. Lenard (2002) demonstrated aggressive corrosion of 70/30 copper–nickel (UNS C71500) specimens exposed to combinations of flowing seawater and sediments rich in decaying organic matter.

Nickel-copper alloy Monel 400 (66.5 percent nickel, 31.5 percent copper, and 1.25 percent iron) (UNS N04400) tubing was severely pitted after exposure to marine and estuarine waters containing SRB (Figure 8-17a, b) (Little et al., 1991). Localized corrosion was attributed to a combination of differential aeration cells, a large cathode-to-anode area ratio, the concentration of chlorides, development of acidity within the pits, and the specific reactions of the base metals with sulfides produced by the SRB. Chloride and sulfur reacted selectively with the iron and nickel in the alloy, leaving a copper-rich, spongy pit interior (selective dealloying) (Figure 8-18). Gouda et al. (1993)

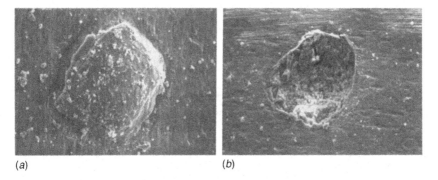

(a)　　　　　　　　　　　(b)

FIGURE 8-17a, b   (a) Blister on the surface of 70/30 nickel-copper (UNS N04400) tube; (b) pit in N04400 nickel-copper tube. (Little et al., 1991.)

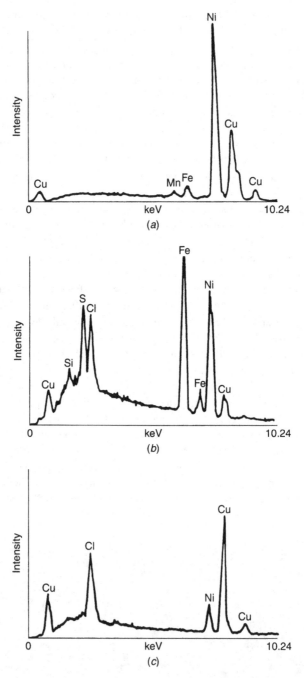

FIGURE 8-18a–c   (a) EDS spectrum of unexposed N04400 nickel-copper alloy. (b) EDS spectrum of nickel alloy after exposure to estuarine water for 6 months showing accumulations of silicon, sulfur, and chlorine with elevated concentrations of iron and nickel. (c) EDS spectrum of the residual metal in the base of the pit showing nickel depletion and copper enrichment. (Little et al., 1991.)

reported that N04400 was highly susceptible to MIC in Arabian Gulf seawater by SRB. The SRB attack was in the form of circular cavities where an intergranular mode of corrosion took place and was accompanied by selective leaching of nickel and iron.

Titanium

No reported cases of MIC of titanium alloys in marine exposures exist; however, titanium surfaces foul quickly and there are reports of plugging and microbial contamination.

SPECIFIC ENVIRONMENTS

Water-Distribution and Storage Systems

Salinity is a measure of salts in water and conductivity is the ability of water to conduct an electrical current. Because dissolved ions increase salinity as well as conductivity, the two parameters are related. In general, in the absence of microorganisms, the rate of corrosion increases with increased salinity/conductivity. Factors such as temperature, pH, redox potential, oxygen concentration, and salt concentration control the particular types of organisms present in water. However, the abundance and availability of an assimilable energy source will determine the extent to which microorganisms impact a habitat. Geesey (1987) states, "Regardless of the effort directed toward maintaining aseptic conditions or sterility, contaminating organisms eventually invade the water system and replicate to the extent the biologically induced problems occur."

Table 8-6 provides a list of water types and their salinity ranges. Dissolved and suspended solids have been removed from distilled water. Deionized water has been through an ion-exchange process in which all charged species or ionizable organic or inorganic salts are removed from solution. Freshwater from wells, rivers, and lakes typically has less than 0.5 parts per thousand (ppt) salinity. Potable water is water that is suitable for drinking, meaning that the water has no impurities present in amounts sufficient to cause disease or harmful physiological effects and conforming in its bacteriological and chemical quality to the requirements of drinking water standards. Potable water is freshwater for which there is no specific salinity

TABLE 8-6 Salinity in Waters

Water types	Salinity (ppt)
Distilled water/deionized	0
Freshwater	<0.5
Brackish water	0.5–17
Seawater	>32

range. Potable water contains numerous types of microorganisms that can cause MIC. Brackish water, found in estuaries where river water mixes with seawater, contains 0.5–17 ppt and seawater 32 ppt or more salinity. Microbiologically influenced corrosion has been reported for 300 series stainless steels maintained or hydrotested with distilled, fresh, brackish, and marine waters. Most of the failures are attributed to underdeposit corrosion due to the formation of tubercles by iron-related bacteria, for example, *Gallionella*, *Leptothrix*, *Crenothrix*, or *Sphaerotilus*. Stoecker and Pope (1993) reported MIC of a S30400 stainless-steel tank in a chemical process plant that had been filled with demineralized water at temperatures between 75 and 90 °C. They concluded that microorganisms had formed deposits on the surface and chlorides had accumulated under the deposits, causing intergranular corrosion and stress corrosion cracking. Elshawesh et al. (1997) reported the failure of S30400 wire and rod screens installed in freshwater wells (500 m deep). The water wells remained stagnant for 2 years. At the time of operation, large quantities of sand and gravel were pumped out with the water. The localized corrosion was pitting near the weld zone and grooves (2–20 mm long). The failures were attributed to underdeposit corrosion caused by SRB and iron-oxidizing bacteria. Borenstein and Lindsay (1993) reported several cases of MIC failure in a stainless-steel cooling water system, comprised of S30400 and S30403, resulting from stagnant untreated well, flood, and brackish waters during hydrotests. Stein (1993) reported MIC in S31600 stainless steel pipes after exposure to Delaware Bay River water. Ray et al. (2002) demonstrated pitting at welds in S31603 stainless steel and Nitronic 50 (UNS S20910) storage tanks exposed to natural seawater after 6 to 10 weeks of exposure to either flowing or stagnant natural seawater. They were able to demonstrate a spatial relationship between surface deposits, bacteria, and pitting (Figures 8-19–8-21).

O'Connor and Banerji (1984) evaluated MIC in potable water distribution systems in the United States. Their study included a national questionnaire of water utilities, an evaluation of five Missouri water supply systems, and a 1-year study of a pipe manifold system to monitor and study water quality changes due to bacteria. The authors maintain that the quality of treated water pumped into water distribution systems is high throughout the United States. However, once the water enters the distribution system changes can occur. The most frequently observed changes include the loss of residual chlorine, decrease in dissolved oxygen, and, in the case of carbon steel, an increase in dissolved iron concentration. Studies with carbon steel, copper, and polyvinyl chloride pipes showed that the carbon steel accumulated the most sediment but the heterotrophic bacteria per unit area of pipe surface was essentially the same for the three materials. In general, higher bacterial counts were observed in locations where consumers reported water-quality problems. A variety of organisms was isolated from five distribution systems: Sulfate-reducing, sulfur-oxidizing, metal-depositing, nitrogen-fixing, nitrifying, and denitrifying bacteria were associated with red and black water. In the case of carbon steel piping, distinguishing between water quality issues and MIC is difficult. Iron- and manganese-depositing bacteria can cause underdeposit corrosion. The accumulation of corrosion products and bacteria causes restrictions of water flow and sloughing results in dirty water—both red and black waters—that have an undesirable taste.

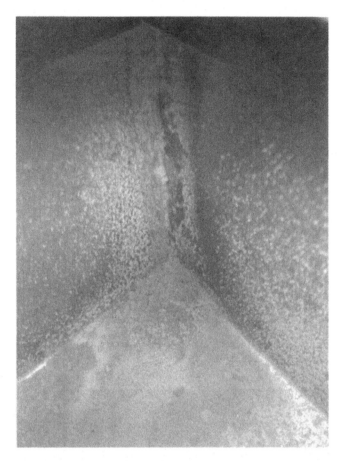

FIGURE 8-19   Corrosion visible at welds on S31603 after a 10-week exposure to flowing seawater. (Ray et al., 2002.)

Each country has their own adopted standard for copper tubing used in water distribution systems. In the U.S., copper pipes conform to ASTM Standard B-88, in most instances, using UNS C10200, C12000, and C12200 alloys. Other country standards include: Britain (BS 2871), France (NF A51-120), and Germany (SF-Cu F38 DIN 1786). In general, the alloys used are greater than 99.0 percent copper. Copper tubing has had a long history of reliable use for water-distribution systems within buildings, however, several types of copper corrosion can occur in potable water systems. These include general corrosion, pitting corrosion, erosion corrosion, and a phenomenon known as "blue water." Reiber (1989) suggested that copper corrosion was most often uniform and pitting occurred only under limited and rare conditions. Uniform corrosion and pitting are often associated with specific water qualities. Hard water contains an appreciable concentration of dissolved minerals. Soft water is treated water that contains sodium as the only cation along with a variety of anions. Uniform corrosion typically occurs when copper is exposed to soft waters of low pH. Pitting failures of copper

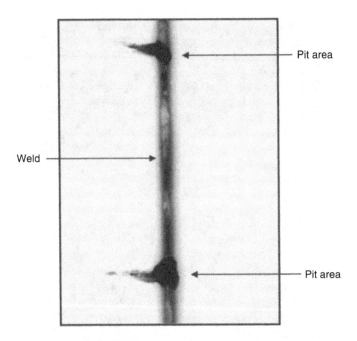

FIGURE 8-20 X-ray image of S31603 stainless-steel weld indicating pit areas after a 10-week exposure to flowing seawater. (Ray et al., 2002.)

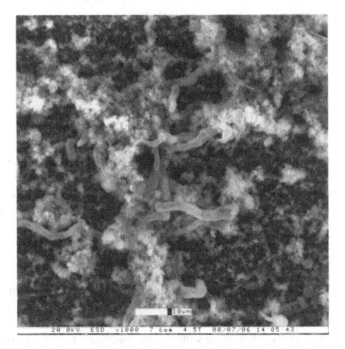

FIGURE 8-21 Biofilm inside a pit on the surface of S31603 coupon after a 10-week exposure to flowing seawater. (Ray et al., 2002.)

tubing has traditionally been categorized as one of three distinct types, that is, Types 1, 2, and 3. More recently, Type 1½ and "pepper-pot" pitting have been identified. "Blue water," Type 1½, and "pepper-pot" pitting have been attributed to MIC. Wall thinning and through-wall penetration has also been attributed to SRB (Labuda, 2003).

Types 1 and 2 pitting were originally classified as "hard" and "soft" water pitting, respectively (Table 8-7). Type 1 pitting results from the presence of an impermeable, usually carbon, film on the inner surface of the pipe. Campbell (1950) demonstrated rapid failure of copper water pipes by pitting corrosion due to carbonaceous films (residual drawing lubricant) formed during pipe manufacture. He determined that the action of the carbonaceous film in producing corrosion of copper tubes was twofold. The film provided a large efficient cathode surface that acted in conjunction with small anodic areas at cracks in the film. The carbon film also prevented oxygen from reaching the copper surface so that protective copper (I) oxide films were not formed in the presence of the carbonaceous film. Campbell (1950) reported that most of the Type 1 pits were in the lower half of the tube. The characteristics of Type 1 pitting are as follows: Well-defined hemispherical pits are formed, which contain soft crystalline copper (I) oxide below a coherent but perforated membrane of copper (I) oxide that is pseudomorphous with the original copper surface. Copper (I) chloride is also present in varying amounts depending on the rate of corrosion. Immediately above the pit is a mound of copper carbonate. Surrounding the pits, there is typically a deposit of calcium carbonate crystals, stained green with corrosion products, overlying a very thin film of carbon, under which a smooth shiny layer of copper (I) oxide is formed. Lucey (1967) showed that the presence of a water-impermeable film on the copper surface would influence the formation of the oxide film and the probability of pitting in hard waters (Figure 8-22).

Type 2 pitting does not involve carbon films and occurs in soft waters above 60 °C with a bicarbonate–sulfate ratio greater than 1 (Table 8-7). Type 2 pitting is characterized by the formation of deep, narrow corrosion pits, which are often branched. The pits contain hard, crystalline copper (I) oxide and are usually covered by small nodules of corrosion product consisting of copper oxides and copper sulfate. The tube

TABLE 8-7 Types of Pitting Failures in Copper Tubing

Type 1	Occurs in cold-water installations of annealed or half-hard copper pipes; it is generally caused by a continuous carbon film on the inside of the pipe formed by cracking of drawing lubricants when the pipes are annealed as a final operation.
Type 2	Found in hot-water installations of hard as well as annealed pipes; it is caused by an unfavorable water composition as will be discussed in Table 8-8.
Type 3	Like type I, it is restricted to cold-water installations, but does not seem to bear any relation to the presence of a continuous carbon film; the main cause is likely to be found in the water composition as explained below.

Source: Mattsson and Fredriksson (1968. Reproduced with permission from Maney Publishing).

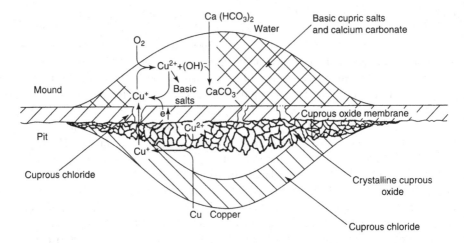

FIGURE 8-22 Diagrammatic representation of the arrangement of corrosion products and the reactions involved in pitting corrosion of copper. (Reproduced from Lucey, 1967, with permission from Maney Publishing.)

surface between the pits is covered with an adherent dull film consisting principally of copper (II) oxide with some copper (I) oxide. A very thin layer of water-deposited silt commonly overlies the whole surface, including the nodules above the pits.

The term Type 3 pitting (von Franque et al., 1975) was used by Linder and Lindman (1983) to describe two cases of pitting occurring in Sweden. The failures occurred in cold-water systems with a high pH, low hardness, and low mineral content. The pitting was characterized by the formation of rather broad areas comprising numerous small hemispherical pits, under a layer of copper sulfate. The surface between the pitted areas was covered with a thin copper (I) oxide film. The pits contained copper (I) oxide with up to 1 percent sulfide.

A type of pitting called "pepper pot" pitting was identified in hospitals in Scotland where the waters were soft, acidic, and had a high humic content from peaty gathering grounds (Keevil et al., 1989). "Pepper pot" pitting produces multiple pits beneath a common copper sulfate crust in both hot- and cold-water systems. Sulfate-reducing bacteria and a variety of aerobes were isolated from the black tubercles covering the pits.

Type 1½ pitting has been reported in institutions in Germany, Saudi Arabia, and Scotland and in private homes in Sweden (Fischer et al., 1995). Wardell and Chamberlain (1994) suggested that a more appropriate name for the phenomenon is "hemispherical MIC pitting." The pitting occurred in hot-, cold-, and warm-water systems and showed some of the characteristics of Type 1 and some of Type 2 pitting. It was therefore termed "Type 1½ pitting." It resembled Type 1 in that the pits were approximately hemispherical and contained soft crystalline copper (I) oxide, with varying amounts of copper (I) chloride, beneath a coherent perforated copper (I) oxide membrane, but resembled Type 2 pitting in that the oxide on the surface between the pits was black and consisted largely of copper (II) oxide. The mounds above the pit were largely of crystalline copper (II) sulfate, often with an outer deposit of powdery copper (II) oxide around the periphery, and on parts of the mound itself. Pitting of the same type has been identified in hospitals in Hellersen, West Germany, Scotland, and England. The involvement of microorganisms with this form of pitting was initially suspected due to the rapid nature of the failures. The first evidence was provided by Fischer et al. (1988), who observed that a gelatinous film could be lifted from the surface of the pipe with the use of 25 percent nitric acid. Chamberlain et al. (1988) demonstrated that gelatinous films removed from corroded copper tubing contained polysaccharides. The polysaccharide films formed by bacteria restricted diffusion to and from the copper surface and facilitated the formation of copper (II) rather than copper (I) oxide. Wardell and Chamberlain (1994) isolated 38 strains of heterotrophic bacteria from standing water and deposits associated with Type 1½ pitting. Scheidt et al. (1999) purified extracellular polymeric substances (EPS) from a biofilm associated with Type 1½ pitting and determined that the material was 95 percent hexose polysaccharides. The following factors delineated by Fischer et al. (1995) contribute to the likelihood of Type 1½ pitting: horizontal orientation, intensively branched piping in large installations, dead ends, long periods of stagnation, and water temperatures between 40 and 45 °C in the warm-water sections. Wagner et al. (1994) summarized the possible corrosion reaction of copper covered with EPS (Figure 8-23). Extracellular polymeric substances are cation-selective

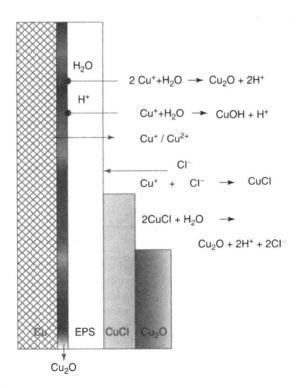

FIGURE 8-23 Possible corrosion reactions on copper covered with biopolymeric material in electrolytes containing chloride. (Wagner, 1994.)

and permeable for copper ions and hydrated protons. Chloride ions cannot diffuse through the EPS layer; therefore the only corrosion reactions at the Cu_2O–EPS boundary are those reactions with water, leading to the formation of hydrated or unhydrated copper (I) oxide or hydroxide. The EPS barrier also inhibits the oxygen reduction reaction. For a review of Type 1½ pitting, see Wagner et al. (1997).

Blue water (also called copper by-product release or cuprosolvency) (Geesey et al., 2002) is sometimes observed in copper tubing, primarily in soft waters after a stagnation period of several hours to days and is typically associated with copper concentrations of 2 to 20 mg L^{-1} (Critchley et al., 2005). Blue water has been reported in New Zealand, Australia, the United States, Japan and Europe (Webster et al., 2000). In all cases, the potable water was a soft, weakly buffered, and slightly alkaline water prepared from a surface water (Fischer et al., 1995) (Table 8-8). Both the cold- and hot-water systems were affected. General attack was detected in all cases. This phenomenon is distinct from other types of copper corrosion in that it does not significantly compromise the integrity of the tube, but instead leads to copper contamination and coloring of the water. This phenomenon received increased attention in the United States after implementation of the U.S. Environmental Protection Agency Lead–Copper rule in 1991 limiting total copper in potable water to a maximum of 1.3 mg L^{-1}. While copper by-product release does not always cause blue water (requires concentrations above 5 mg L^{-1}), it may cause water utilities to exceed the recommended limit. Webster et al. (2000) concluded that the biofilm was important in creating and maintaining a low pH interfacial condition (Figure 8-24a–c). The decreased pH in conjunction with the incorporation of bacterially produced EPS in the copper oxide film decreased the protective nature and created the conditions for blue water.

Within the past decade, the fire-protection sprinkler industry has become increasingly aware of failure incidents due to MIC. Fire-protection systems in commercial and residential buildings are maintained with both fresh and potable waters in carbon steel, galvanized, and copper pipes. Pipe failures include pinhole leaks and blockages in pipes and sprinkler heads. Yee and Whitbeck (2004) used microbial

TABLE 8-8 Water Sources and Water Treatments

Country	Water source	Water treatment
Germany	Storage lake	pH-adjustment: carbon dioxide + lime; disinfection: chlorine + chlorine dioxide
Saudi Arabia	Seawater	Distillation, mixed with well water, partially demineralized by reverse osmosis; disinfection: chlorine
Scotland	Moorish lake	Flocculation, pH adjustment: lime; disinfection: chlorine
Sweden	Mixture of lake and well water	pH-adjustment: carbon dioxide + lime; disinfection: no information

Source: Fischer (1995. © NACE International).

Chapter 6
Major Components of the CT Scanner

.1 System Overview

efore giving a detailed analysis and description of major components in a CT anner, this chapter will present a system overview to explain how the different omponents work together to produce CT images. Figure 6.1 presents a generic lock diagram of a CT system. The actual system architecture for different ommercial scanners may deviate from this diagram, but the general unctionalities of all CT scanners are more or less the same.

For a typical CT operation, an operator positions a patient on the CT table nd prescribes a scanogram or "scout view." The purpose of this scan is to etermine the patient's anatomical landmarks and the exact location and range of T scans. In this scan mode, both the x-ray tube and the detector remain ationary while the patient table travels at a constant speed. The scan is similar a conventional x ray taken either at an A-P position (with the tube located in e 6 or 12 o'clock position) or a lateral position (with the tube located in the 3 or o'clock position). Once such a scan is initiated, an operational control computer istructs the gantry to rotate to the desired orientation as prescribed by the perator. The computer then sends instructions to the patient table, the x-ray eneration system, the x-ray detection system, and the image generation system perform a scan. The table subsequently reaches the starting scan location and aintains a constant speed during the entire scanning process. The high-voltage enerator quickly reaches the desired voltage and keeps both the voltage and the urrent to the x-ray tube at the prescribed level during the scan. The x-ray tube roduces x-ray flux, and the x-ray photons are detected by an x-ray detector to roduce electrical signals. At the same time, the data acquisition system samples e detector outputs at a uniform sampling rate and converts analog signals to igital signals. The sampled data are then sent to the image generation system for rocessing. Typically, the system contains high-speed computers and digital gnal processing (DSP) chips. The acquired data are preprocessed and enhanced efore being sent to the display device for operator viewing and to the data orage device for archiving.

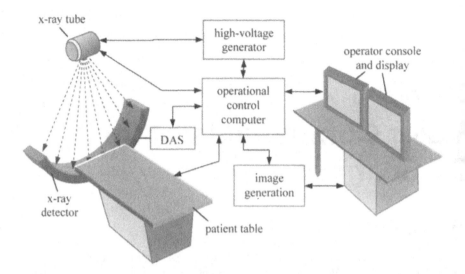

Figure 6.1 Block diagram of a CT system.

Once the precise location and range are determined (based on the scanogram image), the operator prescribes CT scans based either on preset sets of protocol or newly created protocols. These protocols determine the collimator aperture, detector aperture, x-ray tube voltage and current, scan mode, table index speed, gantry speed, reconstruction FOV and kernel, and many other parameters. With the selected scanning protocol, the operational control computer sends a series of commands to the gantry, the x-ray generation system, the table, the x-ray detection system, and the image generation systems in a manner similar to that outlined for the scanogram operation. The major difference between the processes is that the x-ray gantry is no longer stationary. It must reach and maintain a constant rotational speed during the entire operation. Since a CT gantry typically weighs more than several hundred kilograms, it takes time for the gantry to reach stability. Therefore, the gantry is usually one of the first components to respond to the scan command. All of the other operating sequences are similar to the ones described for the scanogram operation.

In many clinical applications, the operational sequence may differ from the one described above. For example, in interventional procedures, the x-ray generation may be triggered by a footpaddle rather than a computer. In contrast-enhanced CT scans, the injection of the contrast agent must be synchronized with the scan, which may require the integration of a power injector with the CT scan protocols. In other operations, the generated x-ray images are sent directly to filming devices to produce hard copies. These deviations, however, should not impact our general understanding of the CT operation mechanism.

6.2 The X-ray Tube and High-Voltage Generator

The x-ray tube is one of the most important components of a CT system. Indeed, x-ray tubes supply the necessary x-ray photons to perform the scan. In the early

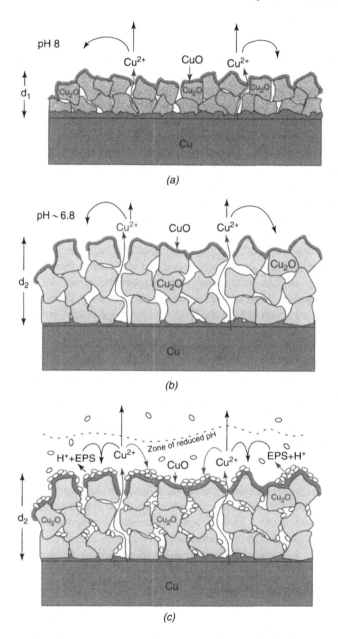

FIGURE 8-24*a–c*   Schematic diagrams indicating the nature of copper oxide films at (*a*) pH 8.0, (*b*) pH 6.8, and (*c*) in instances of MIC. The oxide is thinner and more compact at pH 8.0 compared with the thicker more porous oxide produced at pH 6.8 and instances of MIC. In instances of MIC, the biofilm maintains an interfacial condition of reduced pH and EPS is incorporated into the oxide. (Webster, 2000. © NACE International.)

community analysis, immunoassay analysis, ion chromatography, sulfide testing, and SEM to investigate failed undesignated carbon steel systems (1.5–6-in. diameters) from seven sites in the United States and overseas (Table 8-9). They concluded that the failures were due to MIC but that the microbial populations on the pipe surfaces were very different. Pope and Pope (2000) noted that because water is generally not circulated in fire-protection systems but is regularly added to the front portion of the system, a number of different environments are provided. These environments range from oxygenated with regular nutrient supply to anaerobic environments with few nutrients. Therefore, a single system may have several different degrees of attack. In dry systems, sections of piping that contain some water and trapped air will be vulnerable to MIC. Also, flushing the system with freshwater brings in new bacteria, oxygen, and nutrients. Pope and Pope (2000) list several reasons for the increased reporting of MIC in fire-protection systems: increased awareness, use of thinner-walled pipe, and increased frequency of testing.

The chemical, physical, and microbiological characteristics of industrial waters vary depending on their source, treatment, and application. However, as previously stated, virtually all industrial waters will support the growth of microorganisms. Despite the quantities of freshwater used in food processing, there is a dearth of published information about MIC in the food- and beverage-processing industries. Videla (2001) reported localized corrosion in S31603 stainless-steel plates of a milk sterilizer due to a combination of biological deposits of denatured proteins, high temperatures, and high chlorides present in milk and plant water, but not directly to microorganisms. Elshawesh et al. (2003) described a failure in a 11.4-cm outside diameter, welded, S30400 stainless-steel water pipeline 500 m above grade in a food-processing factory after 3 months of use. The line was supposed to be operated with demineralized water. However, a valve between the well-water tank and the demineralized-water tank leaked and the untreated water (300 mg L^{-1} chloride) remained stagnant in the pipe. The authors concluded that the high chloride concentration initiated the attack and bacteria exacerbated the corrosion process. In both cases, the major cause of the pitting was high chlorides. High-purity waters (distilled/deionized) are used in the pharmaceutical, cosmetic, and electronics industries. Pacheco et al. (1987) reported MIC in undesignated carbon steel cooling water systems in a pharmaceutical plant due to SRB. A sudden increase in the corrosion rate of carbon steel in an open recirculating system from 0.2 to 0.4 mm year^{-1} coincided with an increase in process contaminants.

Nuclear Waste Storage

Microbiologically influenced corrosion in nuclear fuel storage is discussed as a separate topic because it does not fit exclusively into any one environment. Spent nuclear fuels are stored in both aquatic and subterranean environments.

Interim Wet Storage

Storing metal-clad, spent nuclear fuels in water basins pending processing or transfer to interim/long-term dry storage facilities is a worldwide practice that started in

TABLE 8-9 Analysis Summary of Failed Fire-Protection System Samples from Different Locations

Sample set no. and location	Sample description	pH	SRB (bacteria ml^{-1})	SO$_4^{2-}$ (ppm)	Cl^{1-} (ppm)	S^{-2}
1 St. Louis, MO	Source water	9.3	BDL[a]	206	25	Negative
	Sprinkler water with black debris	9.1	1100	103	24	Positive
2 Bedford Heights, OH	Source water	8.1	BDL	29	88	Negative
	Sprinkler water, white/black sludge	8.2	200	29	80	Positive
	Pipe white debris and black sludge	–	200	–	–	Positive
3 Norwood, MA	Source water	7.8	BDL	6	22	Negative
	Sprinkler drop water, white/black debris	7.5	100	37	13	Negative
	Sprinkler drop water, white/black debris, c1965	7.0	100	192	24	Positive
	Sprinkler drop water, white/black debris, c1954	5.8	100	6	9	Negative
4 Melbourne, Australia	Source water	7.5	BDL	24	113	Negative
	Sprinkler water, white/reddish brown debris	7.7	100	13	122	Negative
	Sprinkler water white/brown debris	7.7	1000	0.1	115	Positive
	Sprinkler water white/black debris	7.8	100	0.0	94	Positive
5 Normal, IL	Sprinkler water (NDEP)	8.7	–	22	43	Negative
	Sprinkler water (PHAS)	8.0	BDL	45	57	Negative
	Pipe tubercle/debris	–	BDL	–	–	Negative
6 Harlow, Essex, U.K.	Source water	7.3	BDL	82	38	Negative
	Sprinkler water (side 1) white/rust debris	7.4	1000	5	31	Negative
	Sprinkler water (side 2) white debris	7.6	BDL	52	28	Negative
	Pipe tubercle	–	10000	–	–	Positive
7 San Diego, CA	Source water	7.7	BDL	157	78	Negative
	Sprinkler water	7.5	1000	0.2	92	Negative
	Pipe tubercle	–	10000	–	–	Positive

Source: Yee and Whitbeck (2004. © NACE International).
[a]Below detection limit.

187

mid-1940s. Wet storage uses water as a medium for heat dissipation and radiation shielding. Tables 8-10 and 8-11 summarize wet storage facilities and the materials that are prominent in them (Johnson and Burke, 1997). A world survey of wet storage was conducted under the auspices of the International Atomic Energy Agency (IAEA) in the early 1980s (IAEA, 1982). The survey addressed facilities that were lined with stainless steel and isolated from airborne contaminants, operated with deionized water, involved fuels with high radiation levels, and did not receive fuel from other storage facilities. The response from 115 power reactor storage facilities in 20 countries was that biological impacts were minimal. In 1992, the U.S. Department of Energy began to phase out the reprocessing of spent nuclear fuel. Since that time, both the facilities used for storage and the fuel have aged and the potential for MIC has increased (Wolfram and Dirk, 1997).

Wolfram et al. (1996) conducted a survey of wet storage facilities at Savannah River Site, South Carolina; Oak Ridge National Laboratory, Tennessee; Hanford Site, Washington; and the Idaho National Engineering Laboratory (INEL) (renamed Idaho National Laboratory). They found evidence of biofilms in all basins. However, they concluded that there was insufficient data to determine the importance of MIC in relation to the severity of fuel deterioration, its integrity, or geometry alterations in wet storage, or in transitioning to dry storage. Similarly, Louthan (1997) reported

TABLE 8-10 Overview of Wet Storage Facilities and Characteristics

Features	Power reactor	Defense reactor	Test reactor	Away from reactor[a]
Metal liners	+	−,+	+,−	+,−
Enclosed	+	+,−	+	+
Ion exchange	+	+,−	+,−	+
Receives fuel shipments	−	+,−	−	+

Source: Johnson and Burke (1997). Reprinted with kind permission of Springer Science and Business Media.
[a]Away-from-reactor storage facilities have operated since the 1960s.

TABLE 8-11 Overview of Wet Storage Facilities and Characteristics

Features	Power reactor	Defense reactor	Test reactor	Away from reactor
Stainless steel	+	+	+	+
Aluminum	+,−	+	+	−,+
Carbon steel	−	+	−,+	−
Copper alloys	−	+,−	−,+	−

Source: Johnson and Burke (1997). Reprinted with kind permission of Springer Science and Business Media.

corrosion of fuel cladding, fuel storage racks, and fuel handling/transfer tools at the Savannah River Site but concluded that the probability for MIC was low.

Kopecni et al. (1997) reported corrosion of undesignated aluminum canisters in the spent storage pool of a research reactor in Yugoslavia due to the deterioration of water quality in the pool. Visual inspection showed that all steel construction elements were heavily corroded. Corrosion was also observed on S30400 stainless-steel walls of the basins and transport channels. Sludge has accumulated at the bottom of the pool and radioactivity was high in the pool water. Nykikos et al. (1997) reported localized corrosion, pitting of aluminum-clad fuel rods, in a storage basin in Hungary even where the water was purified at regular intervals. Neither group made reference to specific MIC mechanisms. Filipiak et al. (1997) reported microbial growth and corrosion of irradiated beryllium blocks in a research reactor in Poland. Pitting caused cladding penetration and release of activity from a small fraction of the aluminum spent nuclear fuel and target elements in wet storage.

The Three Mile Island (TMI) Unit 2 reactor accident (near Middletown, PA) produced several types of core debris. During processing of the core debris for storage at INEL, a small amount of ethylene glycol hydraulic fluid was spilled into the core region (Miller et al., 1988). Approximately 2 to 3 months after the cover of the reactor was removed, crews noticed filamentous material hanging on the walls of the vessel (Wolfram and Dink, 1997). All water samples tested positive for microorganisms. Some of the organisms were growing in field radiation close to 10,000 rad h^{-1} in a high-borate (>4000 ppm) and very low-carbon environment (<70 ppm) (Wolfram and Dirk, 1997). Aerobic and anaerobic heterotrophic bacteria and fungal species were identified. Miller et al. (1988) suggested that the hydraulic fluid provided nutrients for microbial growth. Hydrogen peroxide was used to control microbial growth. As canisters were filled, the core debris received an additional peroxide treatment. Despite the precautions, an organism survived and was isolated at INEL. Miller et al. (1988) designed experiments to examine the effect of TMI Unit 2 reactor vessel bacteria on materials used in the construction of TMI canisters [304 stainless steel (UNS S30400), 1100 aluminum (UNS A91100), and an aluminum/boron carbide compact alloy (designated Boral)] designed to contain TMI fuel debris for up to 30 years. They were attempting to evaluate the potential of MIC from within the containers. There was no corrosion of the S30400 stainless, but after 12 months both the A91100 and the aluminum boron carbide alloys were pitted.

Long-term Dry Storage

Most radioactive waste-producing countries plan to dispose high-level waste (HLW) in repositories in subterranean geological formations. Most disposal concepts are based on a series of natural and engineered barriers, designed to prevent or delay the release of radionuclide species and include encasement of the waste in a metal container, placement in a geological formation, backfill with some material (e.g., bentonite), and closure. The performance of a radioactive repository will be determined by the chemistry of both the waste and the materials of construction.

The only absolute barrier for the release of radioactive material will be the container. All other barriers are permeable. Both corrosion-resistant and corrosion-allowance alloys have been considered for container materials. Single- and double-wall designs have been proposed. With corrosion allowance materials (e.g., carbon steel), it is essential to calculate the appropriate thickness to ensure the appropriate container lifetime (Ishikawa et al., 1994). A two-layered undesignated copper–steel canister had been proposed in Sweden. As illustrated in Figure 8-25, the canister would

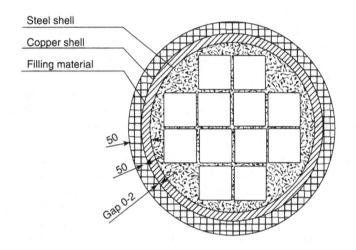

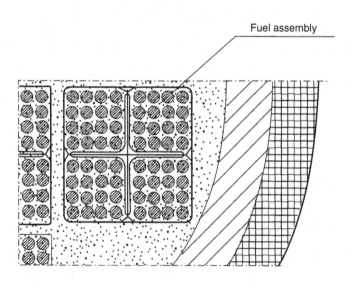

FIGURE 8-25 Schematic design of canister for fuel assemblies together with an enlargement of a fuel assembly to show the configurations of the fuel pins and void spaces. (Reprinted from Oversby and Werme, 1995, with permission from the Material Research Society and the authors.)

consist of an outer layer of about 50-mm-thick copper to provide corrosion resistance and an inner steel layer of similar size to provide structural strength. The canisters would be buried in a mined geologic repository (Oversby and Werme, 1995). In Japan, an isolation system for high-level nuclear wastes has been proposed that would include vitrified waste contained in an overpack, compacted bentonite, and a host rock. Carbon steel is one of the candidate overpack materials (Ishikawa et al., 1994). A concept for permanent disposal of Canada's used nuclear waste was based on containment deep in the stable rock of the Canadian shield. Used fuel bundles would be sealed inside corrosion-resistant containers, surrounded by a compacted buffer (bentonite–sand mixture) and emplaced in a vault excavated 500 to 1000 m deep in the granite rock. A similar scenario has been proposed for nuclear waste repository in Yucca Mount (YM), Nevada.

Bachofen (1991) concluded that microorganisms will be present in all repositories and different habitats will have different location-specific microbial activities. Species diversity is predicted to be quite large in all nuclear waste repositories (Gazso, 1997). In most cases, water availability will be the primary growth-limiting factor. Microorganisms, both indigenous and those introduced through construction activities, could potentially cause MIC.

King and Stroes-Gascoyne (1995, 1997) and Stroes-Gascoyne et al. (1995) evaluated the possibility of MIC of pure C10100 copper and three titanium alloys [UNS R50400 (grade 2), R53400 (grade 12), and R52402 (grade 16)] nuclear fuel disposal containers. They concluded that biofilm formation on the container surface would be unlikely during the initial hot, oxidizing period in the evolution of the vault environment because of the high gamma-radiation fields, elevated temperatures, and desiccation of the environment and suggested that the surface of the containers will be sterilized within 1 to 3 months. They theorized that a zone around the repository would be depleted of microorganisms but over time microorganisms will repopulate the area and could conceivably reach the containers and establish a biofilm on the surface.

Horn et al. (2003) provided a complete roster of microorganisms extant within Yucca Mountain (YM) and a summary of possible MIC mechanisms. Some of the bacteria identified at YM are known to withstand elevated temperatures and extended periods of desiccation. Many are known to maintain activity at low nutrient concentrations. Some of the isolates can degrade complex organic compounds, including substituted aromatics. Others can produce organic acids, enzymes, and extracellular polysaccharides and yet others either consume or degrade nitrates, changing the Cl–anion ratios. Pitonzo et al. (2005) demonstrated that microorganisms isolated from the deep subsurface at YM promoted both uniform and localized corrosion of carbon steel in laboratory experiments. Combinations of bacteria demonstrated higher corrosion rates than individual types. Martin et al. (2004) and Horn et al. (1999, 2000, 2002) evaluated the corrosion behavior of Alloy 22 (UNS N06022), under consideration as the outer barrier candidate for waste packing by exposing to a simulated, saturated repository environment consisting of crushed rock from YM and a continual flow of simulated ground water for periods of up to 5 years. Surface analysis showed development of submicron-sized pinholes and pores. These features were not present on either sterile or untreated control coupons.

Several candidate materials have been suggested for canister-filling material, including glass beads, lead shot, copper spheres, sand, olivine, hematite, magnetite, crushed rock, bentonite clay, other clays, and concrete (Oversby and Werme, 1995). Bentonite (sodium montmorillite), a smectite clay, is stable and has a low water permeability. Philp et al. (1984) suggested that there are strains of SRB that can survive in the clay and the presence of water could exacerbate corrosion of carbon steel. King and Stroes-Gascoyne (1997) used electrochemical experiments and a clay-covered C10100 copper electrode to demonstrate that sulfide ions produced by SRB could diffuse through buffer material and cause corrosion of a copper container. Castro et al. (1998) showed that rock matrix enhanced biofilm formation and the ability of microorganisms to survive at elevated temperatures.

Low- and intermediate-level radioactive wastes (LLW and ILW, respectively) may be encapsulated in bitumen or concrete. The bitumen wastes are then disposed of in deep geological repositories. In Sweden, some LLW and ILW from nuclear power plants will be encapsulated in bitumen, put into steel drums, and placed in rock caverns 50 m below the Baltic Sea. Bitumen is a residue of successive crude-oil treatments. It is a mixture of high-molecular-weight hydrocarbons with such a complex chemical composition that it is undefined. Although undefined, the organic content of bitumen makes it susceptible to biodegradation. After the repository is closed, it will be filled with water. The growth of microorganisms on the surface of bitumen has been demonstrated with electron microscopy. Roffey and Norqvist (1991) concluded that under those conditions, bitumen will be degraded aerobically and anaerobically and that it would be preferable to use a nonbiodegradable material to encapsulate the wastes. Cement has been widely used in the United States as a binder to solidify LLW. Rogers et al. (1997) reported that these waste forms are susceptible to failure in the presence of thiobacilli.

Environments with Hydrocarbons

Microbiologically influenced corrosion due to the microbial contamination and decomposition of hydrocarbons is a persistent problem during production, gathering, transmission, distribution, storage, and use of crude and refined fuels. It is estimated that damage due to MIC in production, transport, and storage of oil amounts to some hundred million dollars in the United States every year (Costerton et al., 1991). Some bacteria and fungi occur naturally in fuel; others are introduced as contaminants from air or water. Microbial interaction with hydrocarbon fuels is limited to water availability. Since water is sparingly soluble in hydrocarbons, microbial growth in hydrocarbons is concentrated at oil–water interfaces, that is, emulsified water, and separate water phases. Walker et al. (1975) compared degradation of hydrocarbons by bacteria and fungi. Bacteria showed decreasing abilities to degrade alkanes with increasing chain length while filamentous fungi did not exhibit a preference for specific chain lengths. The first products of microbial oxidation of hydrocarbons are alcohols, aldehydes, and aliphatic acids. The final products are water and CO_2. The volume of water required for microbial growth in hydrocarbon fuels is extremely small. Since water is a product of the microbial mineralization of organic substrates, it is possible for microbial mineralization of fuel to generate a water phase for further proliferation. For example, *Hormoconis resinae*, the kerosene fungus, grew in 80 mg water per liter of kerosene and after 4 weeks incubation, the concentration of water increased more than tenfold (Bosecker, 1996).

Production

An oil reservoir consists of oil-bearing porous rock, covered by an impervious dome-shaped rock cap. Oil accumulates in the cap, and is prevented from further movement by the impervious rock. Oil production requires drilling a well through the rock and moving the oil up the well tubing. Rate of production depends on the pressure difference between the top and bottom of the reservoir, the permeability of oil-bearing strata, and the pore size in the rock. The pressure difference between the top and the bottom is initially the result of naturally occurring underlying or formation water in the reservoir. Primary recovery is the term used to describe oil recovery using this inherent pressure. Enhanced or secondary recovery is required when the inherent energy is insufficient or has been depleted. Enhanced recovery is achieved by injecting gas or water to the bottom of the reservoir, thereby increasing the pressure difference. In the North Sea, seawater injection is used for secondary recovery. The seawater is filtered and treated with biocides to reduce the possibility of MIC and to prevent the "souring" of the oil by biologically produced hydrogen sulfide (Edyvean and Sneddon, 1986).

The petroleum production environment is particularly suitable for the activities of SRB because it handles large volumes of oxygen-free water from underground reservoirs, which contain the nutrients necessary for their growth (Mora-Mendoza et al., 2001). Microbiologically influenced corrosion of carbon steel used in oil exploration and production due to SRB has been reported around the world. Ciaraldi et al. (1999) concluded that the factors influencing MIC in the Gulf of Suez were low flow velocities, deposit accumulations, waterflooding, and increased levels of bacteria. El-Raghy et al. (1998) reported that undesignated carbon steel pipelines used to produce El-Morgan field crude in the Gulf of Suez lost 75 percent of their original thickness due to the activities of bacteria, particularly SRB. Jenneman et al. (1998) reported MIC in an undesignated carbon steel pipeline used for the disposal of produced water from a coal seam gas field in the San Juan basin. At start-up, the coal seam gas field produces both gas and water. The gas is separated from the water at the surface, is dehydrated, and compressed into a gas gathering line. The coproduced water is either collected in storage tanks and trucked off-site for disposal or pumped into a produced water gathering line for injection into disposal wells. Two leaks were detected in a 4-in. produced water line 25 months after the line was put into service. Both leaks were located in the vicinity of welds and were characterized as underdeposit corrosion due to microorganisms.

Strickland et al. (1996) and Ulman and Kretsinger (1999) evaluated MIC in the Lost Hills Oilfield, Kern County, California. The Lost Hills oil- and water-gathering system is located in a low permeability diatomite formation. Two procedures were undertaken to improve extraction efficiency: intensive hydraulic fracturing and waterflooding. The hydraulic fracturing was accomplished by pumping a viscous polymer at high pressure into the formation. The freshwater used in the waterflooding contained high levels of nitrate. Within 18 months of start-up, the oil–water-gathering system experienced pinhole leaks at the oil–water interface. (corrosion rates 6.8 mm year^{-1}). Acid-producing bacteria and SRB were present in the produced fluids and their numbers were increased by the addition of polymer during the hydraulic fracturing. Within 4 years, the cost of replacing and modifying facilities so that they could be mechanically cleaned was $1.795 millions. The authors attributed the corrosion to the initiation of the extensive fracturing work and the construction of a larger gathering system.

Ciaraldi et al. (1999) described the following typical tactics to prevent MIC in oil production lines and equipment:

- Scale inhibition and paraffin dispersal/dissolution
- Periodic chemical and mechanical cleaning
- Use of fiberglass linings, coatings, and sacrificial anodes in vessel and piping bottoms,
- Reroutings of production to increase flow velocities
- Resizing of needed replacement piping to increase flow velocities
- Rotation of piping to extend life (damaged areas relocated from six-o'clock position
- Refurbishment pigging of pipelines (i.e., multiple pig runs with increased aggressiveness to restore to near bare-metal condition; experience has shown that in many cases 50–100 pig runs are required)
- Routine, periodic, and aggressive maintenance pigging of refurbished pipelines
- Routine, batch biocide treatments (when possible, immediately following pigging, mechanical, or chemical cleaning)
- Improved monitoring with coupons, gas analyses, bacterial culturing of liquid, deposit and pigging debris samples, and increased ultrasonic surveys/smart pig inspections.

Transmission, Distribution, and Storage

Transmission lines deliver natural gas to distant power plants, large industrial customers, and municipalities for further distribution. Petroleum transmission lines deliver crude oil to distant refineries or refined products to distant markets, such as airports or depots, where fuel oils and gasoline are loaded into trucks for local delivery. Because the gas or liquid hydrocarbons have been processed—removing all or most water, MIC attack in transmission pipelines is less common. However, "upsets" in operations can cause water to enter the system.

Distribution lines are a part of natural-gas systems, and consist of main lines that move gas to industrial customers, down to the smaller service lines that connect to businesses and homes throughout a municipality. Table 8-12, provided by the Office of Pipeline Safety, details the accident summary from the 2005 calendar year by commodity. Most accidents (46 of 133) involved crude oil. Furthermore, the accident summary over the same period (Table 8-13) indicates that the major cause of accidents was internal and external corrosion. There are no specific statistics concerning those accidents attributed to MIC, but because most natural-gas distribution lines are carrying dry, processed product with little chance for entry of free water and bacteria, microbial attack of distribution piping is chiefly an external phenomenon.

Most gas and oil transmission and distribution pipelines are carbon steel and most interior surfaces are unprotected. "In practice, it is difficult to ensure that no water enters the transport line with the crude oil" (Whitham, 1995). On the North Sea platforms, oil is stored in subsea concrete cells prior to export to shore. Crude oil and produced water (formation water and injection water) can be transferred to the storage cell. Seawater is used as ballast in the cells and to force oil into transport lines. In other operations, water can enter oil transport lines as an emulsion. When flow velocities are low, the water will form a discrete layer at the bottom of the pipeline (Whitham, 1995). Dias and Bromel (1990) investigated the corrosion of an undersea carbon steel (API grade 5 LX-42, no UNS designation) gas pipeline. The corrosion occurred at low areas where water had accumulated. Analyses demonstrated that pitting occurred under deposits and was due to organic acids produced by bacteria.

Because reservoirs of porous rock at various depths below the surface of the earth where gas or oil was originally produced have exhibited the ability to hold gas under pressure, some of these reservoirs have been used for the underground storage of large quantities of natural gas for future use. The gas is stored in the pore spaces within the rock that previously contained the native gas, oil, or water that was produced from the reservoir. Gas is also stored in water-bearing formations, "aquifer storage," in a few cases where there are no depleted petroleum reservoirs near the market area. Because most of these storage facilities have water that is produced with the gas during withdrawal, the likelihood of internal corrosion is increased, including MIC. In addition to MIC, the activities of SRB and other microorganisms have also cause microbial souring to occur in gas storage fields. Pope and Skultety (1995) reported MIC in a natural-gas storage facility in well tubulers, gathering system pipelines, and processing equipment, for example, separators.

Use

Microbial contamination of fuels in ships and aircraft has been studied for the past 40 years and was studied extensively in the 1980s (Ayllón and Rosales, 1988; Salvarezza et al., 1979). Accumulation of water permits the growth of microorganisms and can lead to a diverse community of bacteria and fungi.

Roffey et al. (1989) documented an increase in the corrosivity of jet fuel (JP4) stored underground in unlined rock caverns in Sweden caused by SRB. Several authors have documented the problem of *Hormoconis resinae* MIC in aircraft fuel tanks. Videla et al. (1993) proposed that microorganisms influenced corrosion of

196

TABLE 8-12 Office of Pipeline Safety Hazardous Liquid Pipeline Accident Summary by Commodity 1/1/2005–12/31/2005

Commodity	No. of accidents	Total accidents (%)	Barrels lost	Property damage ($)	Total damages (%)	Fatalities	Injuries
Anhydrous ammonia	1	0.8	9	170,400	0.3	0	0
Anhydrous ammonia	2	1.5	10	121,957	0.2	0	0
Aviation jet fuel	1	0.8	55	31,489	0	0	0
Bass River crude oil—non-H2s	1	0.8	10,380	247,700	0.4	0	0
Butane/natural gasoline	1	0.8	1	500,000	0.8	0	0
Carbon dioxide	1	0.8	2394	3880	0	0	1
Crude	46	34.6	76,822	13,674,012	20.6	0	0
Crude H1s	1	0.8	3245	210,925	0.3	0	0
Crude oil Heavy Louisiana Sweet (Hls)	1	0.8	0	75,000	0.1	0	0
Crude oil—Heavy Louisiana Sweet	1	0.8	600	3,248,496	4.9	0	0
Crude oil—Hls	1	0.8	50	582,000	0.9	0	0
Crude oil—Sjlb	1	0.8	3393	13,500,550	20.3	0	0
Crude oil-Bcf 17	1	0.8	5232	106,550	0.2	0	0
Diesel	7	5.3	1090	1,007,929	1.5	0	0
Diesel low sulfur	1	0.8	1	80,000	0.1	0	0
Ethane	1	0.8	76	21,742	0	0	0
Ethane/propane mix	2	1.5	15	25,879	0	0	0
Ethane/propane mix	1	0.8	45	178,450	0.3	0	0
Ethane/propane mix	1	0.8	3113	359,361	0.5	0	0
Ethylene	1	0.8	0	1,010	0	0	0
Gasoline	21	15.8	14,966	11,979,288	18	0	1
Gasoline and diesel	1	0.8	40	3,003,124	4.5	0	0

Gasoline UI Reg	1	0.8	225	43,150	0.1	0	0
Gasoline/diesel/jet fuel	1	0.8	100	186,000	0.3	0	0
Isobutane	1	0.8	636	120,000	0.2	0	0
Jet A fuel	1	0.8	126	77,634	0.1	0	0
Jet fuel	2	1.5	2507	279,826	0.4	0	0
Jet fuel turbine	1	0.8	764	3,647,558	5.5	0	0
Low sulfur diesel	1	0.8	900	3,100,000	4.7	0	0
Low sulphur diesel	1	0.8	15	73,910	0.1	0	0
Lube extracted feed (LEF)	1	0.8	662	9550	0	0	0
Mixed petroleum products	1	0.8	3	80,200	0.1	0	0
Natural gas liquids	2	1.5	1709	180,290	0.3	0	0
Ngl liquid	2	1.5	1119	101,977	0.2	0	0
No data	6	4.5	113	2,508,688	3.8	0	0
No. 4/No. 6 oil	1	0.8	2	250,200	0.4	0	0
Normal butane	1	0.8	259	13,932	0	0	0
Propane	8	6	1329	364,782	0.5	1	0
Propylene	1	0.8	1000	175,780	0.3	0	0
Refined liquid Petroleum mixture	1	0.8	2	66,000	0.1	0	0
Transmix	1	0.8	365	250	0	0	0
Unlead gasoline	1	0.8	1145	5,053,065	7.6	0	0
Virgin crude oil	1	0.8	700	15,210	0	0	0
Xylene	1	0.8	250	1,000,000	1.5	0	0
Y-grade	1	0.8	27	0	0	1	0
Total	133	102.0	135,494	66,477,744	100.0	2	2
Average		1019	499,833		2	0

TABLE 8-13 Office of Pipeline Safety Hazardous Liquid Pipeline Accident Summary by Cause 1/1/2005–12/31/2005

Cause	No. of accidents	Total accidents	Barrels lost	Property damages ($)	Total damages	Fatalities	Injuries
Body of pipe	8	6	952	4,052,211	6.1	0	0
Butt weld	3	2.3	126	91,699	0.1	0	0
Car, truck or other vehicle not related to excavation activity	3	2.3	2791	462,757	0.7	0	0
Component	10	7.5	5757	1,178,881	1.8	0	0
Corrosion, external	14	10.5	10,142	5,484,486	8.3	0	1
Corrosion, internal	15	11.3	9678	1,054,008	1.6	0	0
Earth movement	3	2.3	10,302	22,051,550	33.2	0	0
Fillet weld	2	1.5	1021	3,686,782	5.5	0	0
Fire/explosion as primary cause	0	0	0	0	0	0	0
Heavy rains/floods	3	2.3	28,780	1,230,500	1.9	0	0
High winds	5	3.8	24,556	153,500	0.2	0	0
Incorrect operation	8	6	19,010	800,737	1.2	0	0
Joint	2	1.5	203	81,420	0.1	0	0
Lightning	3	2.3	2	1,410,300	2.1	0	0
Malfunction of control/relief equipment	2	1.5	727	13,336	0	0	0
Miscellaneous	10	7.5	1415	706,513	1.1	0	0

Category	n	%	n	$	%			
No data	2	1.5	5	170,149	0.3	0	0	0
Operator excavation damage	3	2.3	1804	172,543	0.3	0	0	0
Pipe seam weld	8	6	10,015	6,817,266	10.3	0	0	0
Rupture of previously damaged pipe	3	2.3	2664	6,923,338	10.4	0	0	0
Ruptured or leaking seal/pump packing	5	3.8	164	202,846	0.3	0	0	0
Temperature	2	1.5	1245	5,098,815	7.7	0	0	0
Third party excavation damage	10	7.5	3025	3,386,365	5.1	0	0	1
Threads stripped, broken pipe coupling	5	3.8	366	191,942	0.3	0	0	0
Unknown	4	3	744	1,055,800	1.6	2	2	0
Vandalism	0	0	0	0	0	0	0	0
Total	133	100.0	135,494	66,477,744	100.0	2	2	0
Average			1019	499,833				

UNS A92024 aluminum alloy fuel tanks by the following mechanisms: (1) removing corrosion inhibitors, including phosphate and nitrate, from the medium; (2) production of corrosive metabolites; (3) establishment of microcenters for galvanic activity, including oxygen concentration cells; and (4) removal of electrons directly from the metals. Several investigators reported a decrease in bulk fuel pH due to metabolites produced during growth of fungi (Rosales, 1985; de Mele et al., 1979; de Schiapparelli and de Meybaum, 1980; de Meybaum and de Schiapparelli, 1980). Until recently, the most common isolate related to aircraft fuel and MIC was the fungus *Cladosporium* (*Hormoconis*) *resinae* (Churchill, 1963; Videla, 1986). de Mele et al. (1979) demonstrated a correlation between growth of *Cladosporium* and pH at fuel–water interfaces and measured pH values between 4.0 and 5.0 in the fuel. Fungus-influenced corrosion has been reported for carbon steel and aluminum alloys exposed to hydrocarbon fuels. Rosales (1985) demonstrated metal-ion binding by fungal mycelia resulting in metal-ion concentration cells on undesignated aluminum surfaces. de Mele et al. (1979) reported that corrosivity increased with contact time due to accumulation of metabolites under microbial colonies attached to metal surfaces. Videla (1996) demonstrated acid-etched traces of fungal mycelia on A92024 aluminum surfaces colonized by *H. resinae* (Figure 8-26). de Meybaum and de Schiapparelli (1980) demonstrated that the metabolic products enhanced aqueous-phase aggressiveness even after the life cycle of *Cladosporium* (*Hormoconis*) was completed. Pitting corrosion was common and severe enough in some cases that aircraft leaked jet fuel while parked on the flight line (Leard et al., 2004).

Smith (1991) suggested that recent changes in the chemical composition of fuel and fuel additives have produced a shift in the microbial community in aircraft fuel from fungi to bacteria. The U.S. Department of Defense investigated the impact of

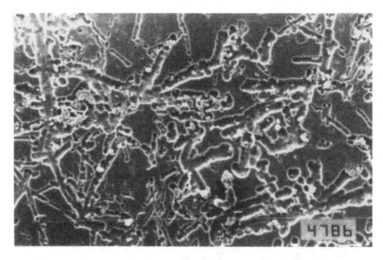

FIGURE 8-26 Scanning electron microscopy micrograph of the etching attack of *H. resinae* on a UNS A92024 aluminum alloy metal surface. Biological and inorganic deposits have been removed by mechanical cleaning (500×). (Reproduced from Videla, 1996, with permission from Routledge/Taylor & Francis Group, LLC.)

fuel additives in 1976 and gradually changed aircraft fuels from jet propellant JP-4 and JP-5 to JP-8 in 1995 (McNamara et al., 2003). JP-8 is kerosene-based with more additives than either JP-4 or JP-5. McNamara et al. (2003) confirmed a shift in the microbial community in aircraft fuel by examining the microbial communities in sump water and biofilms in A92024 aluminum alloy fuel tanks containing JP-8. They found that bacteria isolated from aircraft fuel tanks were closely related to *Bacillus. H. resinae* was not isolated from any of the samples. They further analyzed the microflora to determine the effect of fuel additive diethylene glycol monomethyl ether (DiEGME) on bacterial isolates. McNamara et al. (2005) used EIS (Figure 8-27) and cross-section images (Figure 8-28) to demonstrate that the bacterial isolates could cause MIC of aluminum alloy A92024.

Ships

Hill (1978, 1983) enumerated the following reasons for increased reports of MIC in engines on surface ships: (1) The trend to produce environmentally benign engine oils means that the resulting formulations are more readily biodegraded by bacteria and fungi. (2) Changes in the formulations of lubricating oils have introduced nitrogen, phosphorus, and sulfur required for microbial growth. (3) Low-speed marine engines are at risk because they run for long periods of time at constant temperatures (37–55 °C) conducive to microbial growth. He provided a block diagram of a marine diesel engine indicating the areas susceptible to MIC (Figure 8-29) and a list of oil additives (Table 8-14) that encourage microbial growth.

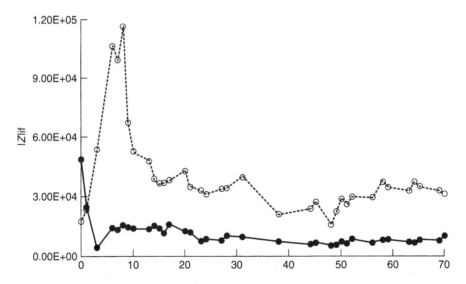

FIGURE 8-27   Low-frequency impedance ($|Z|_{lf}$, 50 mHz) response of A92024 aluminum coupons. Coupons exposed to sterile ASW and JP-8—○; coupons exposed to ASW and JP-8 inoculated with the bacterial consortium—●. (Reprinted from McNamara et al., 2005, with permission from Taylor and Francis Ltd. http://www.tandf.uk/journals.)

FIGURE 8-28   Image of cross-section of an A92024 aluminum coupon (exposed to inoculated JP-8) showing pitting. Scale bar = 400 μm. (Reprinted from McNamara et al., 2005, with permission from Taylor and Francis Ltd. http://www.tandf.uk/journals.)

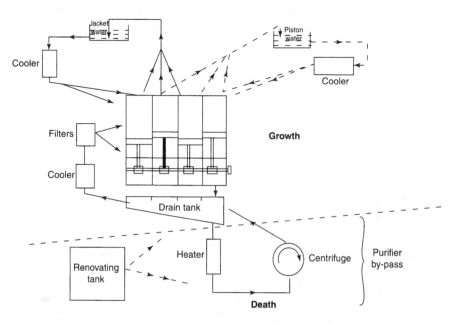

FIGURE 8-29   Block diagram of cross-head marine diesel engine, showing oil and water systems and regions of microbial growth and death. (Reprinted from Hill, 1983, with permission of the National Physical Laboratory, London.)

TABLE 8-14   Typical Oil Additives that Encourage Microbial Growth

Metal soaps, e.g., barium sulfonates
Polyalkenyl succinimides
High-molecular-weight carboxylic acids
Metal dithiophosphates
Polyorganosiloxanes
Hindered phenols, e.g., 2,6-ditertiarybutyl-4-methyl phenol
Aromatic amines—phenyl B naphthylamine
Alkyl phosphates
Alkyl–aryl phosphates, e.g., tricresyl phosphate

Source: Hill (1983). Reprinted with permission of the National Physical Laboratory, London.

Turner et al. (1983) identified three microbial environments in warship fuel tanks: (1) Seawater-displaced fuel tanks may contain a high ratio of water to fuel and a wide range of nutrients in addition to those in the fuel. The nature and quantity of nutrients will depend on the type of water used to fill the header tanks, which may range from saline deep seawater to estuarine water. (2) Undisplaced fuel tanks contain a low ratio of water to fuel and generally less particulate organic matter. Nutrient levels will be determined by the previous history of the fuel and seepage and deterioration of tank linings, seals, and so on. (3) A continuous flow of fuel and water passes over immobilized microorganisms in the coalescer elements.

Microbiologically induced corrosion has been reported in holds, skegs, and oily waste tanks of ships. Huang et al. (1997) reported excessive pitting (up to 2.0 mm year^{-1}) in the uncoated carbon steel (undesignated) bottom plating of cargo oil tanks in crude oil tankers that were 2 to 5 years old (Figure 8-30). The pitting was 4 to 5 times greater than normally anticipated. The investigators determined that a biocide treatment was not prudent because of the cost, lack of efficacy, and safety risks. Instead, they recommended a program of inspections and an epoxy coating. For future designs they recommended better bottom drainage, improved washing of tank bottoms, coatings, and increased thickness of the bottom plate.

Power Generation

Microbiologically induced corrosion has been reported in fossil-fired, nuclear, hydroelectric, and geothermal power plants. The occurrence of MIC in nuclear power plants resulted in an Institute of Nuclear Power Operation Significant Event Report (1984) and a U.S. Nuclear Regulatory Commission (NRC) Information Notice (1985). In 1989, the U.S. Nuclear Regulatory Commission formally recognized the potential for nuclear accidents caused by loss of heat transfer due to microbiological fouling and loss of system integrity caused by MIC (NRC Letter 89-13). Two volumes authored by Pope (1986, 1987) present overview, mechanisms, and case histories of MIC in nuclear and fossil-fired power plants, respectively. The

FIGURE 8-30 MIC pits in uncoated carbon steel (undesignated) cargo oil tanks (5 years old). (Huang, 1997. © NACE International.)

author presented case histories for MIC in carbon steel storage tanks, a carbon steel hot well casing in stainless-steel storage tanks and service water piping, C70600 Cu–Ni fan cooler tubes, N04400 heat exchanger tubes, brass condenser tubes, and aluminum bronze welded pipes. Mechanisms and causative organisms varied with location, susceptible metal, and operating conditions. However, Pope (1987) concluded that in nuclear power plants, MIC was "a problem in most, if not all, systems and alloys (except titanium) used in the industry." Licina (1989) described case histories of MIC in carbon steel (undesignated), austenitic stainless steels (undesignated 300-series), copper alloys (UNS C71590 and C70600), and high-nickel alloys (undesignated 6 percent molybdenum) in safety- and nonsafety-related components/systems in nuclear power plants. He suggested that nuclear power plants have a particular susceptibility to MIC because both long construction times and large standby and redundant systems result in stagnant or intermittent flow in systems exposed to untreated water.

Hayner et al. (1987) found that several welds in the lower half of the condenser water boxes at Crystal River-3 (CR-3), a nuclear power plant located in Citrus County, FL, were seeping seawater. The water boxes were part of a once-through seawater heat exchanger used to condense steam from a low-pressure turbine. CR-3 used four independent systems for this purpose. Each circulated seawater at low pressure and a high flow rate (170,000 gpm). The water boxes were made of S30403 stainless steel. The condenser tubes were ASTM F96 70/30 copper-nickel (UNS C71580). The authors determined that most of the corrosion was caused by MIC assisted by stagnant conditions, internal crevices, and contact with ambient-temperature seawater.

Sinha et al. (1991) reviewed the components, causative organisms, corrosion sites, and failure modes of S30400 and S31600 stainless-steel components in three nuclear power plants—Robinson Nuclear Power Plant, South Carolina; Salem Nuclear Power Plant, New Jersey; and Sequoyah Nuclear Power Plant, Tennessee. The components were located in the reactor coolant, emergency, and reactor auxiliary systems and in the feed water train and condenser. In all cases, the failures were associated with the heat-affected zones of weldments in S30400 and S31600 stainless steel that had been exposed to water from untreated sources during long periods of stagnation. The authors attributed the failures specifically to SRB. However, subsequent investigators have demonstrated that any respiring colony of organisms can cause underdeposit corrosion on these susceptible materials and that sulfide production by SRB is not necessarily the cause of failure.

Brennenstuhl et al. (1991) reported that leaks in iron-nickel-chromium alloy N08800 moderator heat exchanger tubes after a one-year operation at Pickering Nuclear Generating Plant, Ontario, Canada resulted in a shutdown of several units. The purpose of the moderator heat exchanger was to remove excess heat from the heavy-water moderator. The heavy water flows through the nickel horizontal "U" tubes. Untreated freshwater from Lake Ontario is used as the cooling medium on the shell side. The shell side and tube side temperatures range from 3 to 24 °C and 43 to 66 °C, respectively. They demonstrated that pitting in N08800, N08025, and N08028 was due to the formation of a biofilm and the bacteria within the biofilm stabilized sediment particles and caused deposition of calcium carbonate and underdeposit corrosion. Pit density was greatest at the tube sheet and in areas with low flow.

Angell (2002) described the susceptibility of fire-protection systems in nuclear power plants to MIC in Canadian nuclear facilities. The most severe was seen in a plant containing a common ring header fed from four separate high-pressure service water systems. These systems are not balanced and as a result continuously flow within the ring header to feed risers, supplying fire-hose cabinets arranged vertically on the elevations above and below the secondary headers. Each secondary header forms a dead-leg attached to a continuously flowing system. Leaks were located in the risers at the same elevation as the ring header and on the elevation below. Angell (2002) reported that the operation of this system was ideal for iron-oxidizing bacterial growth. With a constant flow of oxygenated water through the ring header, there was a steady supply of oxygen. While the headers and risers were stagnant, the water temperature was ambient. Cold water, rich in both nutrients and oxygen, diffused along the header from the ring main. At the riser cold water mixed with hotter water, which caused the cooler water with the elevated oxygen and nutrients to sink. As this water sank in the riser, it was heated to ambient temperature and eventually rose to be replaced by fresh cooler water entering the system. Heat exchangers that have untreated cooling water on the shell side were found to be susceptible to excessive sedimentation and subsequent MIC.

Rao and Nair (1998) reported the MIC of condenser tube (undesignated admiralty brass) failures at Rajasthan Atomic Power Station, India, after 7 years of operation. The failure of the tubes lead to the leakage of cooling water into the boiler, thereby violating the boiler water technical specifications. The leak rate of cooling water was between 300 and 2100 L h^{-1}. The tubes were damaged due to stress corrosion cracking. They were able to demonstrate that lake water caused the formation of a thick biofilm and that ammonium was produced *in situ* by denitrification. Nitrate reduction produced an average concentration of ammonium exceeding the threshold limit and causing stress-corrosion cracking.

Linhardt and Nichtawitz (2003) reported four cases of corrosion in hydroelectric power plants. The four cases occurred in The Netherlands, Thailand, Illinois, United States, and Austria. In all cases, pitting and crevice corrosion occurred under deposits of manganese oxides on turbine blades made of martensitic stainless steel (UNS J91540) exposed to river water at moderate chloride levels. All cases were identified as MIC due to manganese-oxidizing bacteria. Each unit reported a long period of continuous shutdown.

Torres-Sanchez et al. (1997) reported MIC in the S30403 condenser of a geothermal electric power unit in central Mexico, which operated at temperatures ranging from 150 °C at the inlet to 40 °C in the cooler areas. The authors describe pitting in the tubes as due to SRB.

Paper Mill Industry

The process of making paper is based on the fact that wet cellulose fibers bind together when dried under restraint. The processing of paper making usually involves digestion of a cellulose-containing material, for example, straw, bagasse, cotton, or wood to remove lignin. Caustic soda, sodium sulfate, ammonia, or calcium sulfate can be used to prepare "white waters" for chemical digestions. As the cellulosic material and white water move through the digester, lignin and other components are dissolved, and the cellulose fibers are released as pulp. After leaving the digester, the pulp is rinsed, and the spent chemicals (known as "black water") are separated and recycled. Additional processing is often carried out before or during drying to acquire the desired finish. A typical mill uses multiple stages of bleaching, often with different treatments in each step, to produce a bright white pulp. Once the pulp has been bleached and refined, it is rinsed and diluted with water, and fillers such as clay or chalk may be added. The paper may be coated with starch to improve the printing and strength characteristics.

The paper mill industry requires large quantities of water and stainless steels (S30400 and S31600) are typically used as the construction material. In 1988, the average North American paper mill used 72 m^3 of water for every ton of paper produced. The conditions for microbial growth are extremely favorable in the paper-making process. The pH and temperature (40–46 °C) are optimum for survival and growth. Cellulose, starch, and sugars are carbon sources and casein is a source of nitrogen. Phosphates, nitrates, and sulfates originate from the raw water. Fines, fillers, and other suspended materials provide surface areas for colonization. Many genera of bacteria have been identified in paper mills (Blanco et al., 1996; Harju-Jeanty and Väätänen, 1984; Väisänen et al., 1998; Tatnall, 1981). Carpén et al. (1999, 2001) reported that MIC could occur in pipe systems, storage and feed tanks, adhesive paste lines, and splash zones. Lutey (1993) indicated that more than 30 percent of the maintenance costs in the pulp and paper industry are directly related to corrosion and that "much of the corrosion" could be attributed directly or indirectly to the uncontrolled growth of microorganisms. He further indicated that white water constitutes the single greatest commodity used in paper making and that the entire white-water system is an environment that supports the growth of microorganisms. Tatnall (1981) attributed MIC on S30400 stainless steel to the formation of microbial surface deposits. Linhardt and Mori (2004) detailed the corrosion of S31603 stainless-steel pipes in the process water distribution system of a paper mill due to manganese-oxidizing microorganisms. Preferential attack of the heat-tinted welds was explained by the potential shift caused by biomineralized MnO_2 that prevented repassivation.

REFERENCES

Alanis I, Berardo L, de Cristofaro N, Monia C, Valentini C. (1986). A case of localized corrosion in underground brass pipes. In: Biologically Induced corrosion. *Proceedings of the International Conference on Biologically Induced Corrosion*, June 10–12, 1985, Gaithersburg, Maryland, pp. 102–108.

Angell P (2002). Use of microbial kinetics to control MIC in the nuclear power industry. *CORROSION/2002, Paper No. 02475*. Houston, TX: NACE International, p. 7.

Ayllón ES, Rosales BM (1988). Corrosion of AA 7075 aluminum alloy in media contaminated with *Cladosporium resinae*. *Corrosion*, **44**: 638–643.

Bachofen R (1991). Gas metabolism of microorganisms. *Experientia*, **47**: 508–513.

Barlo TJ, Berry WE (1984). An assessment of the current criteria for cathodic protection of buried steel pipelines. *Mater. Perform.*, **23**(9): 9.

Beech IB, Campbell SA, Walsh FC (1993). Microbiological aspects of the low water corrosion of carbon steel. *12th International Corrosion Congress*, pp: 3735–3746.

Beech IB, Campbell SA, Walsh FC (2001). Marine microbial corrosion. In: Stoecker J (ed) *A Practical Manual on Microbiologically Influenced Corrosion*, Volume 2. Houston, TX: NACE International, pp. 11.3–11.14.

Beech IB, Gillard S, Campbell SA (2004). Accelerated low water corrosion: causes and concerns. *12th International Congress on Marine Corrosion and Fouling*, July 27–30, Southampton, U.K.

Blanco MA, Negro C, Gáspar I, Tijero J (1996). Slime problems in the paper and board industry. *Appl. Microbiol. Biotechnol.*, **46**: 203–208.

Borenstein SW, Lindsay PB (1993). MIC failure analysis. In: Kobrin G (ed) *A Practical Manual on Microbiologically Influenced Corrosion*. Houston, TX: NACE International, pp. 189–194.

Bosecker K (1996). Deterioration of hydrocarbons. In: Heitz E, Flemming H-C, Sands W, Springer V (eds) *Microbially Influenced Corrosion of Materials*, Berlin: Heidelberg, p. 439.

Brennenstuhl AM, Doherty PE, King PJ, Dunstall TG (1991). The effects of biofouling on the corrosion of nickel heat exchanger alloys at Ontario hydro. In: Dowling NJE, Mittleman MW, and Danko JC (eds) *Microbially Influenced Corrosion and Biodeterioration*. Knoxville, TN: The University of Tennessee, pp. 4-25–4-31.

Campbell HS (1950). Pitting corrosion in copper water pipes caused by films of carbonaceous material produced during manufacture. *J. Inst. Metals*, **77**: 345–356.

Carpén L, Raaska L, Hakkarainen T (1999). Effects of biofilm formation on the electrochemical behavior of AISI 304 SS in board machine environment. *CORROSION/1999*, Paper No. 165. Houston, TX: NACE International, p. 19.

Carpén, L, Hakkarainen T, Raaska L, Kujanpää K, Mattila K, Uutela P, Salkinoja-Salonen M (2001). Simulation of MIC at splash zone areas of the paper industry. *CORROSION/2001*, Paper No. 01245. Houston, TX: NACE International, p. 15.

Castro PR, Amy PS, Crossen HV, Jones DA, Southam G, Donald R, Ringelberg DB (1998). The corrosion of carbon steel in rock microcosms containing native Yucca Mountain microorganisms. *CORROSION/1998*, Paper No. 284. Houston, TX: NACE International, p. 26.

Chamberlain AHL, Angell P, Campbell HS (1988). Staining procedures for characterizing biofilms in corrosion investigations. *Br. Corros. J.*, **23**: 197–199.

Chandrasekaran P, Dexter SC (1994). Thermodynamic and kinetic factors influenced by biofilm chemistry prior to passivity breakdown. *CORROSION/1994*, Paper No. 482. Houston, TX: NACE International.

Cheung CWS, Walsh FC, Campbell SA, Chao WT, Beech IB (1994). Microbial contributions to the marine corrosion of steel piling. *Int. Biodeter. Biodegr.*, **34**: 259–274.

Chung Y, Pytlewski KR, McGarry DM (2001). Microbiologically influenced corrosion of TP304L stainless steel underground piping with tape wrapped ER/E316L welds. *CORROSION/2001*, Paper No. 01247. Houston, TX: NACE International, p. 27.

Churchill AV (1963). Microbial fuel tank corrosion: mechanisms and contributory factors. *Mater. Perform.*, **2**(6): 18–23.

Ciaraldi SW, Ghazal HH, Abou Shadey TH, El-Leil HA, El-Raghy SM (1999). Progress in combating microbiologically induced corrosion in oil production. *CORROSION/1999*, Paper No. 181. Houston, TX: NACE International, p. 10.

Copenhagen WJ (1954). Sulphur as a factor in the corrosion of iron and steel structures in the sea. *S. Afr. Ind. Chem.*, **8**: 32–35.

Costerton JW, Boivin J (1991). Biofilms and corrosion in biofouling and biocorrosion. In: Flemming HC, Geesey GG (eds) *Industrial Water System*, Berlin, Heidelberg: Springer, pp. 195–204.

Critchley M, Taylor R, O'Halloran R (2005). Microbial contribution to blue water corrosion. *Mater. Perform.*, **44**(6): 56–59.

Cullimore R, Johnson L (1999). The fate of iron: more lessons from the Titanic tragedy. *Maritime Reporter*, August 1999. p. 22–66.

Cullimore R, Johnson L (2000). The science and the RMS Titanic, the biological odyssey. *Voyage*, **32**: 172–176.

de Mele MFL (1992). Influence of cathodic protection on the initial stages of bacterial fouling. *NSF-CONICET Workshop, Biocorrosion and Biofouling, Metal/Microbe Interactions.* Mar del Plata, Argentina.

de Mele MFL, Salvarezza RC, Videla HA (1979). Microbial contaminants influencing the electrochemical behavior of alumnium and its alloys in fuel/water systems. *Int. Biodeter. Bull.*, **15**(2): 39–44.

de Meybaum BR, de Schiapparelli ER (1980). A corrosion test for determining the quality of maintenance in jet fuel storage. *Mater. Perform.*, **19**(8): 41–44.

de Schiapparelli ER, de Meybaum BR (1980). Microbiological corrosion in terminal storage tanks for aircraft fuel. *Mater. Perform.*, 19(10): 47–50.

Dias OC, Brommel MC (1990). Microbiologically induced organic acid underdeposit attack in a gas pipeline. *Mater. Perform.*, **29**(4): 53–56.

Diercks M, Sand W, Bock E (1991). Microbial corrosion of concrete. *Experietia*, **47**: 514–516.

Dexter SC (ed) (1986). Biologically induced corrosion. *Proceedings of the International Conference on Biologically Induced Corrosion.* Houston, TX: NACE International.

Dexter SC, Gao GY (1988). Effect of seawater biofilms on corrosion potential and oxygen reduction of stainless steel. *Corrosion*, **44**(10): 717–723.

Dexter SC, Lin S-H (1991). Effect of marine bacteria on calcareous deposits. *Mater. Perform.*, **30**(4): 16–21.

Drewello R, Weissmann R (1997). Microbially influenced corrosion of glass. *Appl. Microbiol. Biotechnol.*, **47**: 337–346.

Edyvean RGJ (1984). Interactions between microfouling and the calcareous deposit formed on cathodically protected steel in seawater. *6th International Congress on Marine Corrosion and Fouling*. Athens, Greece, pp. 469–483.

Edyvean RGJ, Sneddon AD, (1986). Biodeterioration problems of seawater injection systems for enhanced oil recovery. In: Barry S, et al. (eds) *Biodeterioration VI*. Slough, UK: CAB International, pp. 497–504.

Ellis SH (1914). Corrosion of steel wharves at Kowloon. *Proceedings of the Institution of Civil Engineering*, **199**(133), Paper No. 4090.

El-Raghy SM, Wood B, Abuleil H, Weare R, Saleh M (1998). Microbiologically influenced corrosion in mature oil fields – a case study in El-Morgan field in the Gulf of Suez. *CORROSION/1998*, Paper No. 279. Houston, TX: NACE International, p. 15.

Elshawesh F, Abusowa K, Mahfud H, El-Agdel E (2003). Microbiologically influenced corrosion of type 304 austenitic stainless steel water pipe. *Mater. Perform.*, **42**(9): 54–60.

Elshawesh F, Elmendelsi T, Elhoud A, Abuan E, Elagdel E (1997). Microbiologically influenced corrosion causes failure of type 304 water well screens. *Mater. Perform.*, **36**(6): 66.

Filipiak B, Hajewska E, Mieleszczenko W, Milczarek JJ (1997). Overview of the interim storage systems for spent fuel at Swierk Centre. In: Wolfram JH, Rogers RD, Gazso LG (eds) *Microbial Degradation Processes in Radioactive Waste Repository and in Nuclear Fuel Storage Areas*. Dordrecht: Kluwer Academic Publishers, pp. 163–170.

Fischer KP (1981). Cathodic protection in saline mud containing sulfate-reducing bacteria. *Mater. Perform.*, **20**(10): 41–46.

Fischer WR, Hanbel PJB, Paradies HH (1988). In: Sequeira CAC, Tiller AK (eds) *Microbial Corrosion 1*. London: Elsevier Applied Science, pp. 300–327.

Fischer WR, Wagner D, Siedlarek H (1995). Microbiologically influenced corrosion in potable water installations – an engineering approach to developing countermeasures. *Mater. Perform.*, **34**(10): 50–54.

Gazso LG(1997). Basic radiation microbiology. In: Wolfram JH, Rogers RD, Gazso LG (eds.) Microbial Degradation Processes in Radioactive Waste Repository and in Nuclear Fuel Storage Areas. Dordrecht: Kluwer Academic Publishers, pp. 93–102.

Gaylarde PM, Gaylarde CC, Guiamet PS, Gomez de Saravia, SG, Videla HA (2001). Biodeterioration of Mayan building at Uxmal and Tulum, Mexico. *Biofouling*, **17**(1): 41–45.

Geesey GG (1987). Survival of microorganisms in low nutrient waters. In: Mittleman MW, Geesy GG (eds) *Biological Fouling of Industrial Water Sysytems*. San Diego, CA: Water Micro Associates, pp. 1–23.

Geesey GG, Baty A, Bremer PJ, Henshaw GS, Schamberger PC, Webster BJ, Wells DB (2002). Chemical characterization of deposits associated with microbiologically influenced copper corrosion in potable water systems. *Proceedings of the Corrosion 2002 Research Topical Symposium*. Houston, TX: NACE International, pp. 1–19.

Gehrke T, Sand W (2003). Interactions between microorganisms and physicochemical factors cause MIC of steel pilings in harbours (ALWC). *CORROSION/2003*, Paper No. 03557. Houston, TX: NACE International.

Gubner R, Beech I (1999a). Statistical assessment of the risk of biocorrosion in tidal waters. *CORROSION/1999*, Paper No. 184. Houston, TX: NACE International.

Gubner R, Beech I (1999b). Statistical assessment of the risk of the accelerated low-water corrosion in the marine environment. *CORROSION/1999*, Paper No. 318. Houston, TX: NACE International.

Gudas JP, Hack AP (1979). Sulfate induced corrosion of copper nickel alloys. *Corrosion*, **35**(2): 67–72.

Guezennec VJ (1991). Influence of cathodic protection of mild steel on the growth of sulfate-reducing bacteria at 35 °C in marine sediments. *Biofouling*, **3**: 339.

Guezennec VJ, Therene M (1988). A study of the influence of cathodic protection on the growth of SRB in marine sediments by electrochemical techniques. In: Sequeira CAC, Tiller AK (eds) *Microbial Corrosion*. London: Elsevier Science, , pp. 256–265.

Gouda VK, Banat IM, Riad WT, Mansour S (1993). Microbiologically influenced corrosion of UNS N0400 in seawater. *Corrosion*, **49**: 63–73.

Hagenauer A, Hilpert R, Hack T (1994). Microbiological investigations of corrosion damages in aircraft. *Werkstoffe und Korrosion*, **45**: 355–360.

Harju-Jeanty P, Väätänen P (1984). Detrimental microorganisms in paper and cardboard mills. *Paperi ja Puu* **3**: 245–251.

Hayner GO, Pope DH, Crane BE (1987). Microbiologically influenced corrosion in condenser water boxes at Crystal River-3. *Proceedings of the Third International Symposium on Environmental Degradation of Materials in Nuclear Power Systems – Water Reactors.* August 30–September 3, Grand Traverse, MI, pp. 647–653.

Hill EC (1978). Microbial degradation of marine lubricants – its detection and control. *Transactions of the Institute of Marine Engineers*, Vol. 90A, pp. 197–215.

Hill EC (1983). Microbial corrosion in ship engines. *Proceedings of the Conference on Microbial Corrosion*, Book 303. March 8–10, Teddington, UK. London: National Physical Laboratory and the Metals Society, pp. 123–127.

Horn J, Carrillo C, Dias V (2003). Comparison of the microbial community composition at Yucca Mountain and laboratory test nuclear repository environments. *CORROSION/2003*, Paper No. 03556. Houston, TX: NACE International, p. 12.

Horn J, Lian T, Martin S (2002). Microbiologically-facilitated effects on the surface composition of alloy 22: a candidate nuclear waste packaging material. *CORROSION/2002*, Paper No. 02448. Houston, TX: NACE International, p. 9.

Horn J, Martin S, Masterson B, Lian T (1999). Biochemical contributions to corrosion of carbon steel and alloy 22 in a continual flow system. *CORROSION/1999*, Paper No. 162. Houston, TX: NACE International.

Horn J, Martin S, Rivera A, Bedrossian P, Lian T. (2000). Potential biogenic corrosion of alloy 22, a candidate nuclear waste packaging material, under simulated repository conditions. *CORROSION/2000*, Paper No. 00387. Houston, TX NACE International, p. 11.

Horvath J, Novak M (1964). Potential/pH equilibrium diagrams of some $Me.S.H_2O$ ternary systems and their interpretation from the point of view of metallic corrosion. *Corrosion Sci.*, **4**: 159.

Huang RT, McFarland BL, Hodgman RZ (1997). Microbial influenced corrosion in cargo oil tanks of crude oil tankers. *CORROSION/1997*, Paper No. 535. Houston, TX: NACE International, p. 21.

International Atomic Energy Agency (IAEA) (1982). *Storage of water reactor spent fuel in water pools*. Technical Report Series 218. Vienna, Austria: International Atomic Energy Agency.

Institute of Nuclear Power Operations (1984). *Microbially Influenced Corrosion (MIC)*. Significant Event Report (SER), pp. 73–84.

Ishikawa H, Honda A, Sasaki N (1994). Long life prediction of carbon steel overpacks for geological isolation of high-level radioactive waste. In: Parkins RN (ed) *Life Prediction of Corrodible Structures*, Vol. I. Houston, TX: NACE International, pp. 454–471.

Ito K, Matsuhashi R, Kato T, Miki O, Kihira H, Wantanabe K, Baker P (2002). Potential ennoblement of stainless steels by marine biofilms and microbial consortia analysis. *CORROSION/2002*, Paper No. 02452. Houston, TX: NACE International.

Jack TR (2001). External corrosion in the gas pipeline industry. In: John Stoecker (ed) *A Practical Manual on Microbiologically Influenced Corrosion*, Vol. 2. Houston, TX: NACE International, , pp: 6.1–6.12.

Jack TR, Francis M McD, Worthingham RG (1986). External corrosion of line pipe part II. Laboratory study of cathodic protection in the presence of sulfate-reducing bacteria.

Proceedings of the International Conference on Biologically Induced Corrosion. Houston, TX: NACE International, , pp. 339–350.

Jenneman GE, Wittenbach P, Thaker JS, Wu Y (1998). MIC in a pipe used for disposal of produced water from a coal seam gas field. *CORROSION/1998*, Paper No. 281. Houston, TX: NACE International.

Johnson R, Bardal E (1985). Cathodic properties of different stainless steels in natural seawater. *Corrosion*, **41**(5): 296–306.

Johnson AB, Burke SP (1997). General and specific perspectives on biological impacts in wet storage facilities for irradiated nuclear fuel. In: Wolfram JH, Rogers RD, Gazso LG (eds) *Microbial Degradation Processes in Radioactive Waste Repository and in Nuclear Fuel Storage Areas*. Dordrecht: Kluwer Academic Publishers, pp. 103–112.

Johnson K, Moulin J, Karius R, Resiak B, Confete M, Chao WT (1993). *Low Water Corrosion on Steel Piles in Marine Waters*. European Commission on Technical Steel Research Final Report, EUR 17868.

Johnson K, Moulin JM, Karius R, Resiak B, Confete M, Chao WT (1994). *Low Water Corrosion on Steel Piles in Marine Waters*. British Steel Technical ECSC Final Report FR/S293-7832.

Kasahara K, Kajiyama F (1986). Role of sulfate reducing bacteria in the localized corrosion of buried pipes. In: Biologically Induced Corrosion. *Proceedings of the International Conference on Biologically Induced Corrosion*. Houston, TX: NACE International, p. 171–183.

Keevil CW, Walker JT, McEvoy J, Colbourne JS (1989). Detection of biofilms associated with pitting corrosion of copper pipework in Scottish hospitals. In: Gaylarde LC, Morton LHE (eds) *Biocorrosion*. Kew, England: Biodeteroration Society, pp. 99–117.

King RA, Skerry BS, Moore DCA, Stott JFD, Dawson JL (1986). Corrosion behavior of ductile and grey iron pipes in environments containing sulphate-reducing bacteria. In: Dexter SC (ed) *Microbially Induced Corrosion*. Houston, TX: NACE International. pp. 83–91.

King F, Stroes-Gascoyne S (1995). Microbially influenced corrosion of nuclear fuel waste disposal containers. *Proceedings of the 1995 International Conference on Microbially Influenced Corrosion*. Houston, TX: NACE International, pp. 35/1–35/14.

King F, Stroes-Gascoyne S (1997). Predicting the effects of microbial activity on the corrosion of copper nuclear waste disposal containers. In: Wolfram JH, Rogers RD, Gazso LG (eds.) *Microbial Degradation Processes in Radioactive Waste Repository and in Nuclear Fuel Storage Areas*. Dordrecht: Kluwer Academic Publishers, pp. 149–162.

Kobrin G (ed) (1993). *A Practical Manual on Microbiologically Influenced Corrosion*, Houston, TX: NACE International.

Kopecni M, Matausek MV, Vukadin Z, Maksin T (1997). Corrosion problems in the research reactor "RA" spent fuel storage pool. In: Wolfram JH, Rogers RD, Gazso LG (eds) Microbial Degradation Processes in Radioactive Waste Repository and in Nuclear Fuel Storage Areas. Dordrecht: Kluwer Academic Publishers, pp. 113–120.

Kumar A, Stephenson LD (2005). Accelerated low water corrosion of steel pilings in seawater. *CORROSION/2005*, Paper No. 05221. Houston, TX: NACE International, p. 13.

Labuda EM (2003). Microbiologically induced corrosion of copper piping system—failure analysis. *CORROSION/2003*, Paper No. 03569. Houston, TX: NACE International, p. 8.

Leard R, Vickers C, Dante J (2004). *Assessment of Integral Fuel Tank Corrosion*. Air Force Research Lab. Evaluation, Report No. AFRL/MLS 04-020 (4349ZTAO/0000).

Lenard DR (2002). The effect of decaying marine organisms on the corrosion of copper nickel alloys in sea water. *CORROSION/2002*, Paper No. 02185. Houston, TX: NACE International.

Li SY, Kim Y, Kho Y, Kang T (2003). Corrosion behavior of carbon steel influenced by sulfate-reducing bacteria in soil environments. *CORROSION/2003*, Paper No. 03549. Houston, TX: NACE International, p. 19.

Licina G. (1989). An overview of microbiologically influenced corrosion in nuclear power plant systems. *Mater. Perform.*, **28**(10): 55–60.

Linder M, Lindman E-K (1983). Investigation of pitting corrosion, type II, in copper pipes. *Proceedings of the 9th Scandinavian Corrosion Congress*, Copenhagen, p. 569–581.

Linhardt P, Mori G (2004). MIC by manganese oxidizers in a paper mill. *CORROSION/2004*, Paper No. 04601. Houston, TX: NACE International, p. 9.

Linhardt P, Nichtawitz A (2003). Microbiologically influenced corrosion in hydroelectric power plants. *CORROSION/2003*, Paper No. 03564. Houston, TX: NACE International, p. 8.

Little BJ (ed) (2002). Microbiologically influenced corrosion. *Proceedings of the CORROSION/2002 Research Topical Symposium*. Houston, TX: NACE International.

Little B, Ray R, Wagner P (2001). Biodegradation of nonmetallic materials. In: Stoecker JG (ed) *A Practical Manual on Microbiologically Influenced Corrosion*, Vol. 2. Houston, TX: NACE International, p. 3.1.

Little BJ, Ray RI, Hart KR, Wagner PA (1995). Fungal-induced corrosion of wire rope. *Mater. Performance*, 34(10): 55–58.

Little B, Ray R, Wagner P, Jones-Mehan J, Lee C, Mansfeld F (1999). Spatial relationships between marine bacterial and localized corrosion. *Biofouling*, **13**(4): 301–321.

Little B, Wagner P, Jacobus J (1988). The impact of sulfate-reducing bacteria on welded copper–nickel seawater piping systems. *Mater. Perform.*, **27**(8): 57–61.

Little BJ, Pope RK, Ray RI (2000) An evaluation of fungal-influenced corrosion of aircraft operating in tropical environments. In: Dean SW, Hernandez-Duque Delgadillo G, Bushman JB (eds) *Marine Corrosion in Tropical Environments, ASTM STP 1399*. West Conshohocken, PA: American Society for Testing and Materials, p. 257.

Little B, Wagner P, Ray R, Jones JM (1991). Microbiologically influenced corrosion of copper alloys in saline waters containing sulfate reducing bacteria. *CORROSION/1991*, Paper No. 101. Houston, TX: NACE International.

Little B, Wagner P, Ray R, McNeil M (1990). Microbiologically influlenced corrosion in copper and nickel seawater piping systems. *Mar. Technol. Soc. J.*, **24**: 10.

Louthan Jr. MR (1997). The potential for microbiologically influenced corrosion in the Savannah River spent fuel storage pools. In: Wolfram JH, Rogers RD, Gazso LG (eds) *Microbial Degradation Processes in Radioactive Waste Repository and in Nuclear Fuel Storage Areas*. Dordrecht: Kluwer Academic Publishers, pp. 131–138.

Lucey VF (1967). Mechanism of pitting corrosion of copper in supply waters. *Br. Corros. J.*, **2**: 175–185.

Lutey RW (1993). MIC in the pulp and paper industry In: Kobrin G (ed) *A Practical manual on Microbiologically Influenced Corrosion*.Houston, TX: NACE International, pp. 25–30.

Mansfeld F, Tsai R, Shih H, Little B, Ray R, Wagner P (1990). Results of exposure of stainless steels and titanium to natural seawater. *CORROSION/1990*, Paper No. 109. Houston, TX: NACE International.

Marsh CP, Beitelman AD, Buchheit RG, Little BJ (2005). *Freshwater Corrosion in the Duluth-Superior Harbor*. Summary of Initial Workshop Findings, U.S. Army Corps of Engineers, ERDC/CERL SR-05-03, p. 28.

Martin S, Horn J, Carrillo C (2004). Micron-scale MIC of alloy 22 after long term incubation in saturated nuclear waste repository microcosms. *CORROSION/2004*, Paper No. 04596. Houston, TX: NACE International, p. 17.

Mattsson E, Fredriksson AM, (1968). Pitting corrosion in copper tubes – cause of corrosion and counter-measures, *Br. Corros. J.*,**3**: 246–257.

Maxwell S (1986). The effect of cathodic protection on the activity of microbial biofilms. *Mater. Perform.*, **25**(11): 53–56.

McNamara CJ, Perry TD, Wolf N, Mitchell R, Leard R, Dante J (2003). Corrosion of aluminum alloy 2024 by jet fuel degrading microorganisms. *CORROSION/2003*, Paper No. 03568. Houston, TX: NACE International, p. 6.

McNamara CJ, Perry IV TD, Leard R, Bearce K, Dante J, Mitchell R (2005). Corrosion of aluminum alloy 2024 by microorganisms isolated from aircraft fuel tanks. *Biofouling*, **221**(5/6): 257–265.

Miller RL, Wolfram JH, Ayers AL (1988). *Studies of Microbially Influenced/Induced Corrosion on TMI-2 Canister D-136B*. DOE Report no. EGG-MS-7744. Idaho National Engineering Laboratory.

Mora-Mendoza JL, Garcia-Esquivel R, Padilla-Viveros AA, Martinez L, Martinez-Bautista M, Angeles-Ch C, Flores O (2001). Study of internal MIC in pipelines of sour gas mixed with formation waters. *CORROSION/2001*, Paper No. 01246. Houston, TX: NACE International.

Morley J, Bruce DW (1983). *Survey of Steel Pilings Performance in Marine Environments*. Corrosion of SSP at Southampton Docks, British Steel Final ECSC Report 7210. KB/804 EUR 8942, EN.

Motoda S, Suzuki Y, Shinohara T, Tsujikawa S (1990). The effect of marine fouling on the ennoblement of alectrode potential for stainless steels. *Corros. Sci.*, **31**: 515–520.

Nekoksa G, Gutherman B (1991). Determination of cathodic protection criteria to control microbially influenced corrosion in power plants. *Proceedings of Microbially Influenced Corrosion and Biodeterioration*. Knoxville, TN: University of Tennessee, p. 6-1.

Nuclear Regulatory Commision (1985). *Microbially Induced Corrosion of Containment Service Water System*. IE Information Notice No. 85-30, United States Nuclear Regulatory Commission Office of Inspection and Enforcement, April 19.

Nykikos T, Pajkossy T, Schiller R (1997). Corrosion in a spent fuel storage basin. In: Wolfram JH, Rogers RD, Gazso LG (eds) Microbial Degradation Processes in Radioactive Waste Repository and in Nuclear Fuel Storage Areas. Dordrecht: Kluwer Academic Publishers, pp. 121–124.

O'Connor JT, Banerji SK (1984) *Biologically Mediated Corrosion and Water Quality Deterioration in Distribution Systems*. Technical Report (U.S. Dept. of Commerce), EPA-600/2-84-056, PB8 4-157494, p. 419.

Ortega-Calvo JJ, Arino X, Hernandez-Marine M, Saiz-Jimenez C (2001) Factors affecting the weathering and colonization of monuments by phototrophic microorganisms. *Sci. Total Environ.*, **167**: 329–341.

Oversby VM, Werme LO (1995). Canister filling materials – design requirements and evaluation of candidate materials. *Materials Research Society Symposium Proceedings*, Vol. 353. Pittsburgh, PA: *Materials Research Society*, pp. 743–742.

Pacheco A, Dishinger TD, Tomlin J (1987). Experiences in controlling microbially-induced corrosion. *CORROSION/1987*, Paper No. 376. Houston, TX: NACE International.

Philp JC, Christofi N, West JM (1984). *The Geomicrobiology of Calcium Montmorillonite (Fuller's Earth)*. FLPU 84-4.

Pitonzo BJ, Castro P, Amy PS, Southham G, Jones DA, Ringelberg D (2005). Microbiologically influenced corrosion capability of bacteria isolated from Yucca Mountain. *Corrosion*, **60**(1): 64–74.

Pintado JL, Montero F (1992). Underground biodeterioration of medium tension electric cables. *Int. Biodeter. Biodegr.*, **29**: 357–365.

Pope DH (1987). *Microbial Corrosion in Fossil-Fired Power Plants – A Study of Microbiologically Influenced Corrosion and a Practical Guide for its Treatment and Prevention*. EPRI Final Report, CS-5495, Project 2300—12 Electric Power Research Institute, Palo Alto, CA.

Pope DH (1986). *A Study of Microbiologically Influenced Corrosion in Nuclear Power Plants and a Practical Guide for Countermeasures.* EPRI Final Report, NP-4582, Project 1166-6, Electric Power Research Institute, Palo Alto, CA, May 1986, p. 72.

Pope DH, Duquette DJ, Johannes AH, Wayner PC (1984). Microbiologically influenced corrosion of industrial alloys. *Mater. Perform.*, **24**(4): 14–18.

Pope DH, Dziewulski D, Frank JR (1991). Case histories of microbiologically influenced corrosion in the gas industry: detection, system analyses and targeted treatment. In: Dowling NJ, Mittelman MW, Danko JC (eds) *Microbially Influenced Corrosion and Biodeterioration.* Knoxville, TN: The University of Tennessee, p. 31.

Pope DH, Morris III, EA (1995). Some experiences with microbiologically influenced corrosion of pipelines. *Mater. Perform.*, **34**(5): 23–28.

Pope DH, Pope RM (2000). Microbiologically influenced corrosion in fire protection sprinkler systems. *CORROSION/2000*, Paper No. 00401. Houston, TX: NACE International, p. 10.

Pope DH, Skultety R (1995). Microbiologically influenced corrosion in natural gas storage fields: diagnosis, monitoring and control. *Proceedings of the International Conference on Microbiologically Influenced Corrosion.* Houston, TX: NACE International, pp. 57/1–57/11.

Rao TS, Nair VR (1998). Microbiologically influenced stress corrosion cracking failure of admiralty brass condenser tubes in a nuclear power plant cooled by freshwater. *Corros. Science*, **40**(11):1821–1836.

Ray RI, Little BJ, Jones-Meehan J (2002). A laboratory evaluation of stainless steels exposed to tap water and seawater. In: Little BJ (ed) *Proceedings of the CORROSION/2002 Research Topical Symposium – Microbiologically Influenced Corrosion*, Houston, TX: NACE International, pp. 133–144.

Reiber SH (1989). Copper plumbing surfaces: an electrochemical study. *J. AWWA*, **81**(7): 114–122.

Ribbe PH (1976). *Sulfide Mineralogy.* CS-58-CS76. Washington, DC: Minerology Society of America.

Roffey R, Edlund A, Norquist A (1989). Studies of jet fuel biodeterioration in rock cavern laboratory model. *Int. Biodeter.*, **25**: 191–195.

Roffey R, Norqvist A (1991). Biodegradation of bitumen used for nuclear waste disposal. In: *Microorganisms in Nuclear Waste Disposal, Part II – A Multi-author Review. Experientia*, **47**: 507–584.

Rogers RD, Hamilton MA, Veeh RH, McConnell Jr. JW (1997). A procedure to evaluate the potential for microbially influenced degradation of cement-solidified low-level radioactive waste forms. In: Wolfram JH, Rogers RD, Gazso LG (eds) Microbial Degradation Processes in Radioactive Waste Repository and in Nuclear Fuel Storage Areas. Dordrecht: Kluwer Academic Publishers, p. 43–54.

Rosales BM (1985). Corrosion measurements for determining the quality of maintenance in jet fuel storage. In: Dexter SC, Videla HA (eds) *Proceedings of the Argentine–USA Workshop on Biodeterioration.* Sao Paulo, Brazil: Aquatec Quimica, p. 135–143.

Rowlands JC (1965). Corrosion of tube and pipe alloys due to polluted sea-water. *J. Appl. Chem.*, **15**: 57.

Saiz-Jimenez C (2001). The biodeterioration of building materials. In: Stoecker JG (ed) *A Practical Manual on Microbiologically Influenced Corrosion*, Vol. II, Chapter 4. Houston, TX: NACE International, pp. 4.1–4.17.

Salvarezza RC, de Mele MFL, Videla HA (1979). The use of pitting potential to satudy the microbial corrosion of 2024 aluminum alloy. *Int. Biodeter. Bull.*, **15**: 125–132.

Sanders PF, Hamilton WA (1986). Biological and corrosion activities of SRB in industrial process plant. In: Dexter SC (ed) *Biologically Induced Corrosion*. Houston, TX: NACE International, p. 47–68.

Scheidt T, Quitschau P, Hinze U, Paradies HH (1999). Purification and structural determination of a native biofilm obtained from copper pipes of a hospital—chemical sequence and partial folding. *Microb. Corrosion*, **29**: 140–169.

Sedricks AJ (1996). Crevice corrosion. *Corrosion of Stainless Steels*, 2nd edn. . New York: Wiley-Interscience, Chapter 5, pp. 176–230.

Sinha UP, Wolfram JH, Rogers RD (1991). Microbially influenced corrosion of stainless steels in nuclear power plants. In: Dowling NJ, Mittelman MW, Danko JC (eds) *Microbially Influenced Corrosion and Biodeterioration*. Knoxville, TN: The University of Tennessee, pp. 4-51–4-59.

Sloan RN (2001). Pipe Coatings. In: Biachetti RL (ed) *Peabody's Control of Pipeline Corrosion*. Houston, TX: NACE International, pp. 7–20.

Smith, R.N. (1991). Developments in fuel microbiology. In: Rossmoore HW (ed) *Biodeterioration and Biodegradation 8, Applied Science*. London: Elsevier, pp. 112–124.

Stein AA (1993). Microbiological corrosion of type 316 stainless steel. In: Kobrin G (ed) *A Practical Manual on Microbiologically Influenced Corrosion*. Houston, TX: NACE International, pp. 175–182.

Stoecker JG (ed) (2001). *A Practical Manual on Microbiologically Influenced Corrosion*, 2nd edn. Houston, TX: NACE International, Vol. 2.

Stoecker JG, Pope DH (1993). MIC in metalworking processes and hydraulic systems. In: Kobrin G (ed) *A Practical Manual on Microbiologically Influenced Corrosion*. Houston, TX: NACE International, , pp. 175–182.

Stranger-Johannessen M, Norgaard E (1991). Deterioration of anti-corrosive paints by extracellular microbial products. *Int. Biodeter.*, **27**: 157–162.

Stranger-Johannessen M (1984). Fungal corrosion of the steel interior of a ship's hold. *Biodeterioration VI, Proceedings*. Slough, U.K.: CAB International Mycological Institute, pp. 218–223.

Strickland LN, Fortnum RT, Du Bose BW (1996). A case history of microbiologically influenced corrosion in the Lost Hills Oilfield, Kern County, California. *CORROSION/1996*, Paper No. 297. Houston, TX: NACE International, p. 20.

Stroes-Gascoyne S, Lucht LM, Borsa J, Delanet TL, Haveman SA, Hamon CJ (1995). Radiation resistance of the natural microbial population in buffer materials. *Materials Research Society Symposium Proceedings*. Pittsburgh, PA, Vol. 353, pp. 345–352.

Tatnall RE. (1981). Case histories: bacteria-induced corrosion. *CORROSION/1981*, Paper No. 130. Houston, TX: NACE International, p. 15.

Torres-Sanchez R, Magana-Vazquez A, Sanchez Yanez JM, Gomez LM (1997). High temperature microbial corrosion in the condenser of a geothermal electric power unit. *Mater. Perform.*, **36**(3): 43–46.

Trick KA, Keil G (1999). Fungi and bacterial degradation of polyamide coated aircraft material. *CORROSION/1999*, Paper No. 167. Houston, TX: NACE International.

Turner APF, Eaton RA, Jones EBG (1983). Nutritional aspects of ship fuel system contamination by *Cladosporium resinae*. *Biodeterioration*, **5**: 507–516.

Ulanovskii IB, Ledenev AV (1981). *Influence of Sulfate-Reducing Bacteria on Cathodic Protection of Stainless Steels* (translated from *Zashchita Metallov*, **17**(2)). London: Plenum Press, p. 202.

Ulman MAS, Kretsinger MB (1999). A continued case history of microbial influences in the Lost Hills Oilfield, Kern County, California. *CORROSION/1999*, Paper No. 160. Houston, TX: NACE International, p. 6.

Väisänen OM, Weber A, Bennasar A, Rainey FA, Busse H-J, Salkinoja-Salonen MS (1998). Microbial communities of printing paper machines. *J. Appl. Microbiol.*, **84**: 1069–1084.

Videla H (2001). Localized corrosion of stainless steel milk sterilizers. In: Stoecker JG (ed) *A Practical Manual on Microbiologically Influenced Corrosion*. Houston, TX: NACE International, Vol 2, pp. 11.35–11.38

Videla HA (1996). *Manual of Biocorrosion*. New York: CRC Lewis Publishers, p. 42.

Videla HA (1986). The actions of *Cladosporium resinae* growth on the electrochemical behavior of aluminum. In: Dexter SC (ed) *Biologically Induced Corrosion*, Houston, TX: NACE International, pp. 215–222.

Videla HA, Guiamet PS, DoValle S, Reinoso EH (1993). Effects of fungal and bacterial contaminants of kerosene fuels on the corrosion of storage and distribution systems. In: Kobrin G (ed) *A Practical Manual on Microbiologically Influenced Corrosion*. Houston, TX: NACE International, p. 125.

Videla HA, Guiamet PS, Gomez de Saravia S, Herrera LK, Arroyave C, Poire DG (2003). Assessment of microbiological and atmospheric effects on rock decay. *CORROSION/2003*, Paper No. 03571. Houston, TX: NACE International.

von Franque O, Berth D, Winkler B (1975). Ergebnisse von untersuchungen an deckschichten in kupferrohren. *Werkst. Korros.*, **26**: 4.

von Wolzogen Kuhr CAH, van der Vlugt LS (1934). De grafiteering van gietijzer als electrobiochemisch proces in anaerobe grunden. *Water* (den Haag), **18**: 147–151.

Wagner D, Chamberlain AHL, Fischer WR, Wardell JN, Sequeira CAC (1997). Microbiologically influenced corrosion of copper in potable water installations—a European project review. *Mater. Corrosion*, **48**: 311–321.

Wagner D, Siedlarek H, Fischer WR, Wardell JN, Chamberlain AHL (1994). Corrosion behaviour of biopolymer modified copper electrodes. *Microb. Corrosion*, **15**: 85–104.

Walker JD, Austin HF, Colwell RR (1975). Utilization of mixed hydrocarbon substrate by petroleum-degrading microorganisms. *J. Gen. Appl. Microbiol.*, **21**: 27–39.

Wardell JN, Chamberlain AHL (1994). Bacteria associated with MIC of copper: characterization and extracellular polymer production. *Microb. Corrosion*, **15**: 49–63.

Webster BJ, Werner SE, Wells DB, Bremer PJ (2000) Microbiologically influenced corrosion of copper in potable water systems—pH effects. *Corrosion*, **56**(9): 942–950.

Whitham TS (1995). Challenges to the prediction and monitoring of microbiologically influenced corrosion in the oil industry. In: Tiller A K and Sequeira C A C (eds) *Microbial Corrosion: Proceedings of the 3rd International European Federation of Corrosion Workshop*. London, U.K.: The Institute of Materials, pp. 314–321.

Wolfram JH, Dirk WJ (1997). Biofilm development and the survival of microorganisms in water systems of nuclear reactors and spent fuel pools. In: Wolfram JH, Rogers RD, Gazso LG (eds) Microbial Degradation Processes in Radioactive Waste Repository and in Nuclear Fuel Storage Areas. Dordrecht: Kluwer Academic Publishers, pp. 139–148.

Wolfram JH, Mizin RE, Jex R, Nelson L, Garcia KM (1996). *The Impact of Microbially Influenced Corrosion on Spent Nuclear Fuel and Storage Life*. DOE Report INEL – 96/0335, Idaho National Engineering Laboratory.

Yee GG, Whitbeck MR (2004). A microbiologically influenced corrosion study in fire protection systems. *CORROSION/2004*, Paper No. 04602. Houston, TX: NACE International.

Zhang H-J, Dexter SC (1995). Effect of marine biofilms on initiation time of crevice corrosion for stainless steels. *CORROSION/1995*, Paper No. 285, Houston, TX: NACE International.

Chapter 9

Microbiologically Influenced Corrosion of Nonmetallics

INTRODUCTION

The following sections present a review of biodegradation of nonmetallic materials, including polymeric materials, reinforced polymeric composites, concrete, asphalt and wood. Because of their chemistries these material are susceptible to very specific types of microbial degradation. Biodegradation of building materials was addressed in Chapter 8.

POLYMERIC MATERIALS

Until recently, little attention has been paid to environmental degradation of polymeric materials. It is now recognized that polymeric materials are subject to biofouling and biodeterioration, resulting in loss of flexibility, elasticity, and strength due to fracture, disbonding, or delamination. Possible mechanisms for microbial degradation of polymeric materials include direct attack by acids or enzymes, blistering due to gas evolution, enhanced cracking due to calcareous deposits and gas evolution, and polymer destabilization by concentrated chlorides and sulfides. Polymeric materials are also subject to degradation from moisture intrusion and osmotic blistering. Both may be influenced by biofilms. Organic additives including plasticizers, fillers, and stabilizers, many of the ester type, may provide nutrients for microbial growth and ultimate degradation.

Küster (1979) and Upsher (1976) prepared tables (Tables 9-1 and 9-2, respectively) summarizing the resistance of some polymeric materials to biodegradation. There are conflicts between the data presented in the two tables. For example, Upsher (1976) indicated that polyamides were resistant and Küster (1979) indicated that the resistance of polyamides to microbial attack was low. Because of the range

Microbiologically Influenced Corrosion By Brenda J. Little and Jason S. Lee
Published 2007 by John Wiley & Sons, Inc.

TABLE 9-1   Application and Microbial Resistance of Synthetic Polymers

Polymeric substance	Application	Microbial resistance
Polyethylene	Packaging films, insulation, containers	Very high
Polypropylene		Very high
Polyvinyl chloride	Packaging films, foam	Very high
Polyvinylidene chloride	Packaging films of high chemical stability	High
Polyvinyl acetate	Packaging films, varnish fabrics	Moderate
Polyvinyl alcohol	Packaging films	High
Polyvinyl butyral		High
Polystyrene	Films, foam (Styropor)	High
Polymethyl emthacrylate	Plexiglas	High
Polyacrylonitrile	Fabrics (Orlon, Dralon)	
Polytetrafluroethylene	Insulation (Teflon)	High
Polytrifluorochloroethylene	Insulation (Hostaflon)	High
Cellulose acetate	Acetate rayon	High
Cellulose nitrate	Celluloid	No resistance
Polyamides	Fabrics (Perlon, Nylon)	Low
Polyethylene terephthalate	Fabrics (Terylene, Diolen)	Fair
Polyester resins	Mixed with glass fibers	
Polycarbonates	Utensils	
Silicone	Coatings	High
Phenolformaldehyde	Bakelite	High
Ureaformaldehyde		High
Melaminformadlehyde		No resistance

Source: Küster (1979). Reprinted with permission of John Wiley & Sons, Inc.

of chemical compositions of the polymers, the classifications are not absolute. Furthermore, structural modifications that make polymers resistant to degradation by fungi are not necessarily effective in imparting resistance to bacteria, and vice versa (Stahl and Pessen, 1953).

Biomedical Applications

Medical-grade silicone is used in tubing, catheters, mammary implants, plastic reconstruction, encapsulation of electric components, and voice prostheses (Neu et al., 1993). van der Mei et al. (1996) evaluated the biodeterioration of silicone rubber used for voice prostheses. They observed that silicone rubber voice prostheses in patients after total laryngectomy become rapidly colonized by yeasts and bacteria. Colonies of yeasts grew into the rubber after 3 to 4 months, causing blisters and

TABLE 9-2 Susceptibility of Polymers to Microbial Attack

Susceptible	Variable	Resistant
Polyester polyurethanes	Formaldehyde resins	Polyolefins
Naturally occurring polymers	Substituted celluloses	Substituted polyolefins
		Polyamides
	Polyesters	Polyvinyl resins
	Polyether polyurethanes	Acrylics
		Synthetic rubbers
		Polycarbonates
		Expoxy resins
		Polystyrene
		Polysulfides
		Silicones

Source: Upsher (1976). Permission granted by Prof. Helmut Hagel.

holes (Figure 9-1*a–d*). The authors speculated that the mechanism for the degradation may have been a combination of mechanical disruption and chemical attack due to the release of extracellular enzymes or free radicals. Gettleman et al. (1983) reported that silicone rubber dental liners could be degraded by yeasts.

Polymeric Coatings

Moisture- and chemical-resistant polymeric coatings and linings are used to protect underlying metals against corrosion. Coating performance is influenced by composition, thickness, continuity, adhesion to the metal substratum, and resistance to microbial degradation. Microbial production of acids and enzymes may result in selective leaching of coating components with increased ion transport and porosity.

Dittmer et al. (1975) evaluated the corrosion of iron pipes, encapsulated by a polyethylene film, exposed to sulfate-reducing bacteria (SRB). They reported that polyethylene was susceptible to microbial breakdown. The coatings were permeable to water and soluble sulfide. With increasing time, the corrosion rate and weight losses of the coated specimens were similar to the uncoated specimens. Jack et al. (1996) evaluated coatings in biologically active soils. They found that wet, clay-rich soils fostered higher populations of bacteria than moist sandy soils and polyethylene tape supported higher counts of bacteria than extruded polyethylene or fusion-bonded epoxy coatings, presumably due to the presence of degradable adhesive primer components in the tape. Walch and Jones-Meehan (1990) demonstrated microbial breaching of epoxy and nylon coatings on 4140 steel when the coupons were exposed to mixed communities of marine microorganisms. Gu et al. (1994) studied fungal degradation of polyimide films used as insulaors in electronic packaging. Increased ionic permeability was observed using electrochemical impedance spectroscopy (EIS) after one week's exposure to a fungal consortium, in which

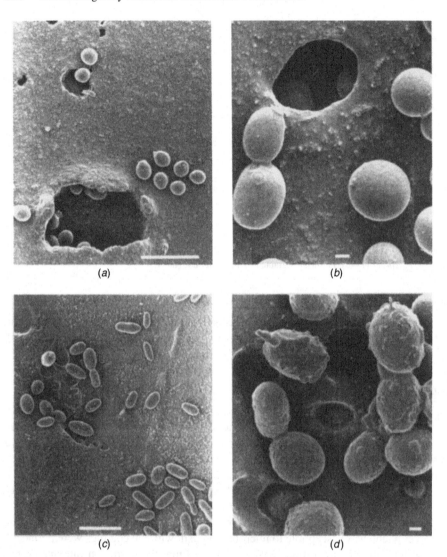

(a)   (b)

(c)   (d)

FIGURE 9-1a–d   Hole-like defects in silicone rubber created by various yeast strains in a modified Robbins device after a 14-day cycle of feast and famine. *C. albicans* strains GB 1/2 (*a*) and GB 9/4 (*b*). *C. tropicalis* strains GB 9/9 (*c*) and GB 19/4 (*d*). The bars equal 10 μm (*a, c*) and 1 μm (*b, d*). (Reprinted from van der Mei, 1996, with permission from the author.)

Aspergillus versicolor was the dominant species. Resistance of the films dropped from 10^8 to 10^6 Ω, indicating significant changes in the coating dielectric properties. Kay et al. (1993) demonstrated reduction in tensile strength and percentage elongation at break for polyester polyurethanes exposed to bacteria. Stranger-Johannessen (1986) confirmed fungal degradation of polyurethane cable sheathing in the marine environment. Polyester polyurethanes and other polyesters are readily broken down by microbial esterases. The resistance of polyester polyurethanes depends on the type of diol. Low-molecular-weight, unbranched alkane diols tend to be susceptible.

Spatial relationships between microorganisms and blisters in coatings have received considerable attention. In humid atmospheric exposures, Stranger-Johannessen and Norgaard (1991) observed that microbial metabolites attacked the coating surface and changed chemical and physical properties without coating penetration. In their studies, blister–substratum interfaces were usually dry, and when fluid was present it was often sterile or contained microorganisms that would not induce coating degradation. Breaches in the coatings expose metal substrata to further localized attack. Jones et al. (1991) described breaches in nylon, epoxy, and polyurethane coatings over UNS G41400 steel exposed to mixed marine communities of anaerobic and facultative bacteria, including SRB. Additionally, epoxy, but not nylon, coatings showed pinpoint holes, blistering, and peeling after a one-month exposure to a nonmarine bacterial assemblage including SRB. Blisters in the coating were typically associated with microorganisms (Figure 9-2a–c). The authors suggested that microorganisms responsible for breaching the coatings may not necessarily be those that initiate localized corrosion on the bare metal under the

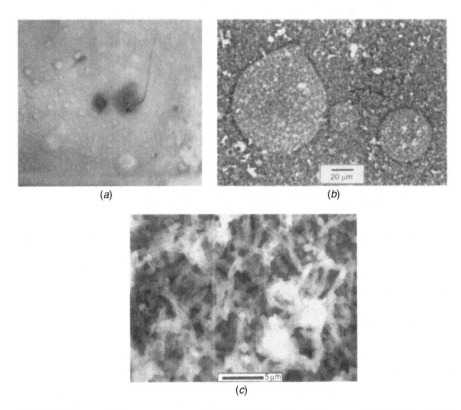

(a)   (b)

(c)

FIGURE 9-2a–c   Blisters on epoxy-coated UNS G41400 carbon steel coupon (a, b) with associated bacteria (c). (Jones et al., 1990.)

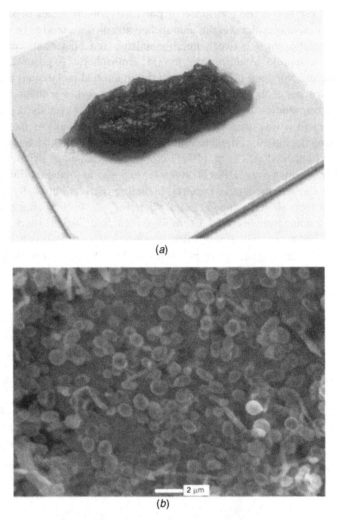

(a)

(b)

FIGURE 9-3a, b   (a) Conductive caulk applied to UNS G41400 steel after a 15-month exposure to a mixed, SRB-containing culture and (b) micrograph of microbial biofilm under the caulk on the steel surface. (Jones-Meehan et al., 1994.)

failed coating. Jones-Meehan et al. (1994) demonstrated water and bacterial intrusion under a polythioether conductive caulk that resulted in corrosion of underlying steel (Figure 9-3a, b). Silicone sealants were resistant to microbial attack.

Fiber-Reinforced Polymeric Composites

Fiber-reinforced polymeric composite (FRPC) materials include a polymer matrix or resin with fiber reinforcement. Fibers, usually glass or carbon, provide strength with low density. Resin matrices provide load transmission and energy dissipation.

Fiber–resin interface bonding is essential for composite integrity. Fiber-reinforced polymeric composites are commonly used in atmospheric and aquatic environments to replace conventional materials in structural applications. They have been used in marine heat exchangers, condensers, pumps, shipboard structures, and deep-sea applications. Fiberglass-reinforced plastics are used as piping and pipe linings and as protective claddings over metals and concrete. Performance advantages include increased strength-to-weight ratio, hardness, wear and corrosion resistance, stiffness, and improved creep behavior. Modifications during fabrication, including thermal properties and configurations, can be tailored for specific applications. Fibers, resins, and the fiber–resin interface have differing susceptibilities to biodegradation.

In laboratory experiments (Wagner et al., 1994, 1996), bacterial cultures were chosen to represent specific degradation mechanisms for worst-case exposures of FRPC coupons. Cultures included *Thiobacillus ferroxidans*, a sulfur/iron oxidizing bacterium; *Pseudomonas fluorescens*, a calcareous-depositing bacterium originally isolated from polluted seawater; *Lactococcus lactis* subsp. *lactis*, ATCC #19435, an ammonium-producing bacterium; *Clostridium acetobutylicum*, ATCC #824, a bacterium previously shown to produce copious amounts of hydrogen from fermentation of sugars; and a mixed facultative/anaerobic marine culture containing SRB originally isolated from corroded carbon steel in marine service. Two FRPCs—a carbon fiber (T-300)-reinforced epoxy and a glass (S-2)- and carbon (T-300)-reinforced vinyl ester—were exposed to all microbiological cultures for 161 days. Additionally, carbon fibers, glass fibers, vinyl ester, and epoxy resins were individually exposed for 90 days to SRB and hydrogen-producing bacteria. Glass fibers had been treated with an organofunctional silane sizing surfactant to promote adhesion between the resin and fibers and facilitate handling.

Surfaces were uniformly colonized by all physiological types of bacteria; however, bacteria preferentially colonized surface anomalies including scratches and fiber disruptions (Figure 9-4*a–c*). Epoxy and vinyl ester neat resins, carbon fibers, and epoxy composites were not adversely affected. Neither the epoxy nor the vinyl ester composites was adversely affected by calcareous-depositing or ammonium-producing bacteria. There was no evidence of attack of resins, and fibers remained embedded within both matrices. Composites exposed to sulfur/iron-oxidizing bacteria were covered with crystalline deposits containing iron and sulfur in addition to microbial cells. All surfaces exposed to SRB were black due to the deposition of iron sulfides. No damage to the epoxy composite, epoxy neat resin, or carbon fibers could be attributed to the presence and activities of SRB and hydrogen-producing bacteria. Hydrogen-producing bacteria appeared to disrupt bonding between fibers and vinyl ester resin (Figure 9-5). Disruption may have been due to gas formation within the composite. Glass fibers (Figure 9-6*a*) exposed to SRB lost all rigidity after a 90-day exposure so that the weave pattern was no longer evident (Figure 9-6*b*). Control glass fibers remained rigid and maintained the original weave pattern after exposure to sterile culture medium (Figure 9-6*c*). There was no direct attack of the glass.

The impact of SRB on the tensile strength of stressed FRPC was evaluated using acoustic emission (AE). Flat, rectangular lengths of a carbon fiber-reinforced epoxy resin composite (6 ply, unidirectional fiber volume 50 percent) were prepared and

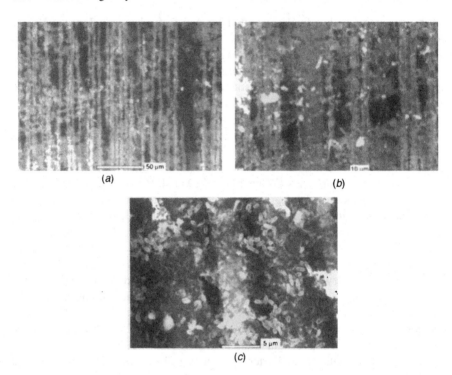

FIGURE 9-4a–c Bacteria and crystals preferentially distributed along fiber/resin interface. (Wagner et al., 1996.)

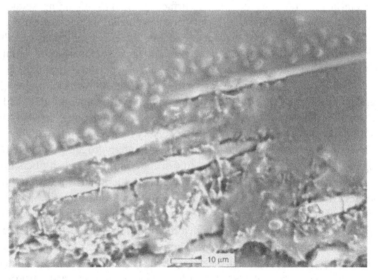

FIGURE 9-5 Hydrogen-producing bacteria at disrupted interfaces between fibers and vinyl ester resin. (Wagner et al., 1994.)

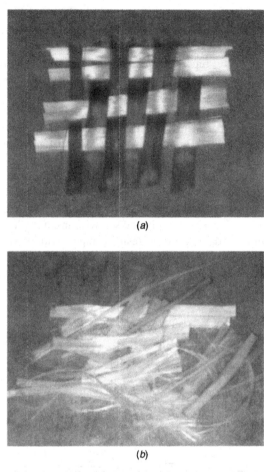

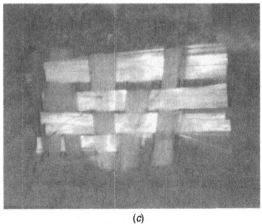

FIGURE 9-6a–c Glass fibers (2X): (a) unexposed, (b) exposed to SRB in culture medium, and (c) exposed to sterile culture medium. (Wagner et al., 1994.)

assembled in three- and four-point bends at controlled strains. Acoustic emission testing suggested a 10-percent loss in tensile strength as the result of microbial exposures (Figure 9-7a, b).

Gu et al. (1995, 1996) reported fungal degradation of carbon-reinforced epoxy, carbon-reinforced bismaleimide, and glass-reinforced fluorinated polyimide composites due to hyphae penetration into resin interiors. Electrochemical impedance spectroscopy was used to demonstrate increasing permeability over 8 months of exposure to a naturally occurring fungal culture containing *Aspergillus versicolor* and a *Chaetomium* species. The authors concluded that microorganisms derived energy from resins and fiber sizings.

The emphasis of this section has been on the negative aspects of biodegradation of polymeric materials. However, biodegradation can have positive effects. Synthetic polymeric composites of polyactide, polyglycolide, and polyhydroxybutyric acid are designed for use as implanted screws, pins, and plates to fix fractured bones in place. Biodegradable polymer resins compete with plastic materials in the form of mulch films, planting containers, hay twine, surgical stitching, medicine capsules, and composting bags (Beach et al., 1996). Biodegradable conformal

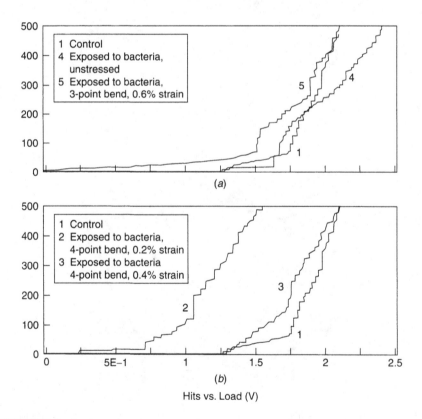

FIGURE 9-7a, b   Acoustic emission plots of hits versus load (V). (Wagner et al., 1996.)

coatings for printed boards and components can be removed by microorganisms rather than toxic solvents and detergents. Biosynthetic materials, such as polyhydroxyalk kanvates, poly-L-motale, and polycaprolactone, are candidate materials to replace thermoplastics because of their biodegradability, especially in seawater (Molitoris, 1995).

CONCRETE

Concrete is an inert aggregate, such as rock and gravel, surrounded by a cement binder. Concrete is a moderately porous mixture of alkaline inorganic precipitates and mineral aggregates, primarily hydrated calcium silicate and portlandite $(Ca(OH)_2)$. Concrete is used in worldwide structural applications due to its strength, durability, and low cost. It is used for large storage tanks and pipes, interior linings, and exterior ballast coatings for steel pipes and containers, paving, and for many other applications. A major construction application is municipal wastewater collection and treatment systems. Concrete is one proposed barrier for nuclear waste storage in subterranean geological formations. Deterioration of concrete will result when any environmental agent can break the inorganic bonds of the cement binder. Acids, sulfates, ammonium and magnesium salts, alkalies, organic esters, and carbon dioxide can destroy a binder over time.

Microbiologically influenced corrosion in concrete sewers has been described by investigators around the world (Islander et al., 1991; Iwamatsu et al., 1988; Sand and Bock, 1984; Sand et al., 1987, 1992). Severyn (1991) estimates that the deterioration of concrete in wastewater collection and treatment systems is primarily due to biodegradation. Sewers are typically built to last 80 to 100 years. Mori et al. (1991, 1992) determined a corrosion rate of 4.3 to 4.7 mm per year in sewer pipes in Japan, or an approximate 20-year lifetime. Sydney et al. (1996) estimated that 10 percent of the sewer pipes in Los Angeles County have been subjected to MIC. Cost estimates for replacement/repair range from several hundred million to one billion dollars (Mansfeld et al., 1991). The overall cost of restoration of sewer networks in Germany damaged by MIC was $1.1 million (Kaempfer and Berndt, 1998). Estimates for sewer repair in Flanders were $5 million per year (Vincke et al., 2000). Similar problems have been reported in the Arabian Gulf region, which are enhanced by elevated temperatures and a scarcity of flush water (Saricimen et al., 1987). The probability of MIC increases with increased organics in the sewage, increased temperature, increased sludge, decreased oxygen, decreased flow rate, and decreased drop along the pipe.

Kulpa and Baker (1991) and Mansfeld et al. (1990, 1991) describe the complex biodegradation process in sewage collection systems involving aerobic and anaerobic bacterial mediation of the sulfur cycle (Figure 9-8). Under anaerobic conditions, SRB reduce sulfate to sulfide in the sewage. Hydrogen sulfide (H_2S) is released into the aerobic environment at the sewer pipe crown above the sewage where it is oxidized to sulfuric acid by sulfur-oxidizing bacteria or thiobacilli. Dissolution of the concrete pipe occurs due to the corrosive action of the acid.

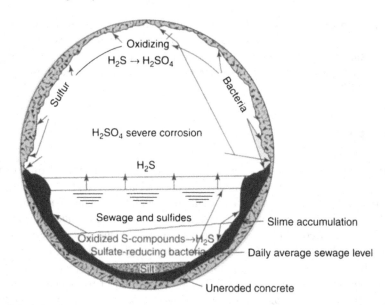

FIGURE 9-8 A schematic representation of the sulfur cycle occurring in sewage pipes. (Vincke, 2000. Reprinted from *Environmental Technologies to Treat Sulfur Pollution*, with permission from the copyright holders, IWA.)

Heterotrophic microorganisms can also induce degradation of cement by excreting carboxylic acids during the decomposition of organic matter (Lea, 1970). Maruthamuthu et al. (1997) demonstrated that heterotrophic bacteria adversely affect the compression strength of concrete. Perfettini et al. (1991) evaluated degradation of cement binder as induced by metabolic products of two fungal strains, *Aspergillus niger* and *Mycelia sterila*, isolated from soil samples. Portland cement [15 percent (w/w) portlandite ($Ca(OH)_2$)] was exposed in direct contact to the two fungal strains for 11 months. Acids produced by *A. niger* (gluconic and oxalic) dissolved cement, increasing porosity by 11.4 percent and reducing the bending strength (strength applied to three points on the surface that causes breakage) by 78 percent. Increased porosity indicates cement dissolution and permits penetration of biological and chemical species. *Mycelia sterila* (gluconic and malic acids) caused a significant 4.2-percent leaching of original calcium content, an 11-percent increase in porosity, and a 62-percent loss in bending strength. Major organic acids produced by fungi solubilize calcium, silica, aluminum, and iron minerals. Solubilization is related more to the nature of the acid than to concentration.

Tator (2003) suggested four strategies to minimize MIC in sewer systems: (1) better design, (2) better materials, (3) modification of the sewage environment, and (4) installation of a barrier between the sewage and the cementitious materials. The author suggested that rapid flow would decrease the time for biological activity and that pipes operating at full capacity with little or no void space would prevent oxidation of the H_2S. He suggested that vitrified clay and plastic piping would be immune to attack by H_2SO_4. Several investigators, in addition to Tator (2003), have suggested that MIC of concrete could be controlled by decreasing sulfate concentration in sewage by a membrane or ion-exchange process or by chemical oxidation or precipitation using ferric or ferrous chloride, hydrogen peroxide, or sodium nitrate (Sydney et al., 1996). Ozonation, air injection, addition of biocides, addition of metals (e.g., selenium), and an alkaline crown spray to maintain pH have also been suggested. Daczko et al. (1997) evaluated the degradation of concrete in mixtures of sulfuric acid. They added metakaolin, silica fume, and an organic corrosion inhibitor (Tables 9-3 and 9-4). They were able to demonstrate that addition of silica fume and organic corrosion inhibitor reduced degradation. Acid-resistant linings and coatings have also been designed to prevent MIC.

Microbiologically influenced corrosion has also been detected in specific types of concrete construction, which provide environments for the growth of microorganisms. Davies and Scott (1996) reported MIC of support columns for a building in Buffalo, New York. The columns were H-shaped, ranging in flange thickness from 0.495 to 1.12 in. (12.6–28.4 mm) and web thickness between 0.31 and 0.68 in. (7.9–17 mm) (Figure 9-9). The columns were originally surrounded by a wet vermiculite/cement fireproofing mix. The corrosion was detected 3 years after construction as blisters on the sheathing of the columns. The corrosion was attributed to the presence and activities of thiobacilli that lowered the pH of the fireproofing mixture and dissolved the concrete. *Desulfovibrio*, an SRB, was also isolated from the anaerobic, low-pH area.

Most concrete structures include carbon steel reinforcements, commonly referred to as rebar, that will readily corrode if exposed as a result of concrete deterioration (Figure 9-10a, b) (Little et al., 2001). Mittleman and Danko (1995) determined that sulfate reduction and sulfide oxidation by microorganisms were responsible for concrete and carbon steel (undesignated) deterioration in a dam in South America. Moosavi et al. (1986) reported that the biogenic production of H_2S by SRB could permeate concrete and convert the passive iron oxide film on undesignated carbon steel rebars into a nonprotective iron sulfide film.

Posttensioning is a method of reinforcing concrete with high-strength steel strands that keep the concrete in compression. Posttensioning applications include

TABLE 9-3   Mixture Proportions of Concrete

Cement	390 kg m^{-3} (658 lb yd^{-3})	Type 1, Meeting ASTM C 15015
Sand	712 kg m^{-3} (1200 lb yd^{-3})	Natural concrete sand
Stone	1068 kg m^{-3} (1800 lb yd^{-3})	#57 Gradation limestone
Water	148 kg m^{-3} (250 lb yd^{-3})	
HRWR	520–780 mL per 100 kg (8–12 fl oz cwt^{-1a})	Napthalene sulfonate based
Slump	75–125 mm (3-5 in.)	
Air	5–8%	

Source: Daczko (1997. © NACE International).
[a]cwt = 100 lb of cement.

TABLE 9-4 Admixtures Used in the Evaluation of Concrete Degradation due to Sulfuric Acid

Mixture number	Admixture	Dosage	General description/effect on concrete
1	–	–	Plain or reference
2	Metakaolin	8% cement replacement by mass	Clay: densifies and reduces permeability
3	Silica fume	8% cement replacement by mass	Pozzolan: densifies and reduces permeability
4[a]	OCI	650 mL hwt^{-1} (10 fl oz cwt^{-1})	Ester/amine chemicals: reduces permeability
5[a]	OCI'	1300 mL hwt^{-1} (20 fl oz cwt^{-1})	Ester/amine chemicals: reduces permeability

Source: Daczko (1997. © NACE International).
Note: All materials were supplied by Master Builders Inc.
[a]OCI—organic corrosion inhibiting admixture designed to protect steel reinforcement via an amine film and reduce permeability via ester pore lining technology. OCI and OCI' are different only in terms of dosage. Dosage of OCI is half that recommended for this admixture.

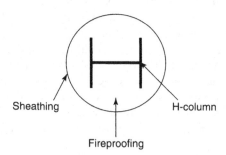

FIGURE 9-9 Schematic cross section of column with associated sheathing and fireproofing. (Davies and Scott, 1996. © NACE International.)

office and apartment buildings, parking structures, slabs-on-ground, bridges, sports stadia, and nuclear power plants. A failure analysis was conducted to determine the cause of a failure of a seven-strand carbon steel (undesignated) cable used as a tendon in posttension construction of a commercial building (Little et al., 2001). Each cable was made of six strands wrapped around a central core (Figure 9-11). Carbon steel

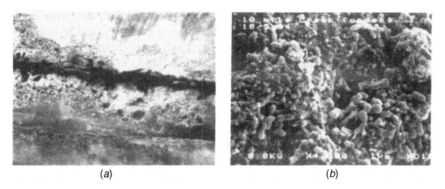

FIGURE 9-10a, b (a) Corrosion of undesignated carbon steel reinforcing bar in concrete and (b) bacteria associated with corrosion product. (Little et al., 2001.)

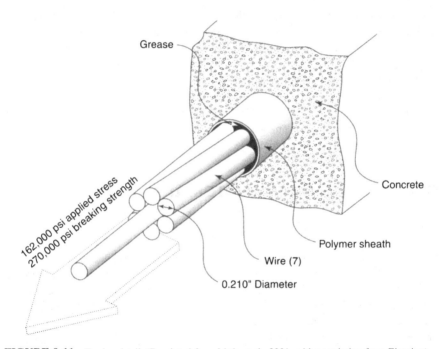

FIGURE 9-11 Tendon detail. (Reprinted from Little et al., 2001, with permission from Elsevier.)

cables used as tendons are typically lubricated with hydrocarbon grease before insertion into preplaced polyvinyl chloride ducts in concrete slabs. Polyvinyl chloride ducts provide corrosion protection and the grease facilitates insertion of the cable into the duct. Cables are posttensioned from one or both ends after the concrete has achieved sufficient strength and anchor plates are attached. Water can be introduced during construction or can accumulate after construction. Little et al. (2001) isolated *Fusarium* sp., *Penicillium* sp., and *Hormoconis* sp. from corroding tendons in a posttensioned structure. The pH of the water associated with the corrosion products was consistently 2 or below. The isolates were used to inoculate tendons in sheathes. In laboratory experiments, Little et al. (2001) demonstrated grease degradation and concomitant acid production using Fourier-transform infrared spectroscopy (FTIR).

MIC has been documented for prestressed tendons in a concrete reactor vessel at Fort St. Vrain Generating Station, Denver, CO (Ashar et al., 1994; Nuclear Regulatory Commission, 1984). After an investigation using metallographic examination, grease analysis, scanning electron microscopy, and biological analysis, investigators concluded that microbiological breakdown of organic grease resulted in the formation of formic and acetic acids, which combined with moisture and caused corrosion. Corrosion was observed in areas where grease had been consumed or removed during placement of the tendons. Causative organisms were not specifically identified.

Similarly, MIC was identified as the cause of breaks in tensioned cables in a silo built of Portland cement in Thayngen, Germany (Richner, 1996). Cables were single strand (diameter 15 mm), coated with lithium 12-hydrostearate grease in a polyethylene tube (diameter 20 mm, wall thickness > 1 mm). Failures, due to reduction of cable diameter, occurred in areas where cables entered anchor plates and between anchor plates and sheathing. In all cases, there was a visible alteration of the condition of the grease in association with corrosion products. Corrosion products ranged from weakly acid to neutral. The watery extract had a vinegary smell and acetic acid was identified in corrosion products.

OTHER ENGINEERING MATERIALS

Asphalt

Biodegradation of asphalt cement-20 by aerobic bacteria has been investigated by Pendrys (1989). Asphalt is the unrefined residue of the fractional distillation of petroleum crude oil and is used extensively as cement in road construction. Petroleum-degrading bacteria excrete biosurfactants that emulsify petroleum hydrocarbons by reducing interfacial tension between the hydrocarbons and water. Pendrys (1989) exposed asphalt for 3 weeks to a mixed culture (*Pseudomonas*, *Acinetobacter*, *Alcaligenes*, *Flavimonas*, and *Flavobacterium* genera) cultivated to use asphalt as a sole carbon source. The author concluded that resultant brittleness and flaking were limited mainly to the asphalt surface. The saturate and naphthalene aromatic distillation phases supported increased bacterial growth with concomitant emulsification.

Wood

Wood is one of the most commonly used engineering materials, with its main components being cellulose, polyoses (polysaccharides), and lignin. Cellulose cell structure determines tensile strength, polyoses influence wood swelling and shrinkage, and lignin makes the wood resistant to pressure (Institute of Materials, 1992). Fungal species are the most common microorganisms in colonization/attack, especially under humid conditions. Bacterial attack on wood is known but is not as common. Bluestain fungi (*Ascomycetes* and *Deuteromycetes*) and molds (*Zygomycetes*, *Ascomycetes*, and *Fungi imperfecti*) colonize the wood surface, causing staining and disfigurement but not loss of strength. Fungal colonization of coatings on painted and processed woods may cause ingress of moisture. *Basidiomycetes* soft rot fungi (brown and white) may, however, degrade one or more of all three main wood components, leading to loss of strength and weight in addition to disfigurement.

SUMMARY

Most nonmetallic engineering materials are ultimately susceptible to microbial degradation. To a limited extent, susceptibility to biodegradation can be predicted from the chemistry of the material. For example, polymeric bonds that can be attacked by enzymes are vulnerable to microbial degradation. The alkaline nature of concrete makes it vulnerable to attack by microbiologically produced acids. The rate of degradation will determine the usefulness of a material in an engineering application. For example, laboratory evidence indicates that asphalt is biodegradable at an extremely slow rate. However, there are no known case histories that document the problem in an engineering application.

REFERENCES

Ashar H, Tan CP, Naus D (1994). Prestressed concrete in US nuclear power plants (Part 2). *Concrete Int.* **16**(6): 5861.

Beach DE, Boyd R, Uri ND (1996). Expanding biodegradable polymer resin use: assessing the aggregate impact on the U.S. economy. *Appl. Math. Model*, **20**: 388–398.

Daczko JA, Johnson DA, Amey SL (1997). Decreasing concrete sewer pipe degradation using admixtures. *Mater. Perform.*, **36**(1): 51–56.

Davies M, Scott PJB (1996). Remedial treatment of an occuppied building affected by microbiologically influenced corrosion. *Mater. Perform.*, **35**(5): 54–57.

Dittmer CK, King RA, Miller JDA (1975). Bacterial corrosion of iron encapsulated in polyethylene film. *Br. Corros. J.*, **10**(1): 47–51.

Gettleman L, Fischer DJ, Farris C (1983). Self-sanitizing soft denture liners: paradoxical results. *J. Biomed. Mater. Res.*, **17**(4): 731–734.

Gu J-D, Ford TE, Thorp KEG, Mitchell R (1994). Microbial degradation of polymeric materials. *Proceedings of the Tri-Service Conference on Corrosion*, Orlando, FL, p. 291.

Gu JD, Ford TE, Thorp KEG, Mitchell R (1995). Microbial deterioration of fiber reinforced composite materials. In: Angell P, Borenstein SW, Buchanan RA, Dexter SC, Dowling NJE, Little BJ, Lundin CD, McNeil MB, Pope DH, Tatnall RE, White DC, Ziegenfuss HG (eds) *1995 International Conference on Microbially Influenced Corrosion*. Houston, TX: NACE International, pp. 25-1-25-5.

Gu JD, Lu C, Thorp K, Crasto A, Mitchell R (1996). Fungal degradation of fiber-reinforced composite materials. *CORROSION/1996*, Paper No. 275. Houston, TX: NACE International.

The Institute of Materials (1992). *Microbiological Degradation of Materials and Methods of Protection*. London, UK: The Institute of Materials, European Federation of Corrosion Publications, No. 9, p. 37.

Islander RL, Devinny JS, Mansfeld F, Postyn A, Shih H (1991). Microbial ecology of crown corrosion in sewers. *J. Environ. Eng.*, **117**(6): 751–770.

Iwamatsu J, Nishizaki K, Katano Y, Horino K (1988). Internal corrosion in sewers caused by hydrogen sulfide. *Corros. Eng.*, **37**: 221–228.

Jack TR, Van Boven G, Wilmot M, Worthingham RG (1996). Evaluating performance of coatings exposed to biologically active soils. *MP*, **35**(3): 39–45.

Jones JM, Walch M, Mansfeld FB (1991). Microbial and electrochemical studies of coated steel exposed to mixed microbial communities. *CORROSION/1991*, Paper No. 108. Houston, TX: NACE International.

Jones-Meehan J, Vasanth KL, Conrad RK, Little B, Ray R (1994). Corrosion resistance of several conductive caulks and sealants from marine field tests and laboratory studies with marine, mixed communities containing sulfate-reducing bacteria. *International Symposium on Microbiologically Influenced Corrosion (MIC) Testing*. Miami, FL: American Society for Testing and Materials.

Kaempfer W, Berndt M (1998). Polymer modified mortar with high resistance to acid and to corrosion by biogenous sulfuric acid. *Proceedings of the IXth ICPIC Congress*, Bologna, Italy, pp. 681–687.

Kay MJ, McCabe RW, Morton LHG (1993). Chemical and physical changes occurring in polyester polyurethane during biodegradation. *Int. Biodeter. Biodegr.*, **31**: 209–225.

Kulpa CF, Baker CJ (1991). Involvement of sulfur-oxidizing bacteria in concrete deterioration. In: Dowling NJ, Mittleman MW, Danko JC (eds) *Microbially Influenced Corrosion and Biodeterioration*. Knoxville, TN: University of Tennessee, pp. 4/7–4/10.

Küster E (1979). Biological degradation of synthetic polymers. *J. Appl Polym. Sci.*, **25**: 395–404.

Lea FM (1970). *The Chemistry of Cement and Concrete*, 3rd edn. London: Arnold.

Little BJ, Ray RI, Wagner PA (1993). Biodegredation of non-metallic engineering materials. In: Stoecker JG (ed) *A Practical Manual on Microbiologically Influenced Corrosion, Vol. 2*. Houston, TX: NACE International, pp. 3–7.

Little BJ, Staehle R, Davis R (2001). Fungal influenced corrosion of post-tensioned cables. *Biodeter. Biodegr.*, **47**: 71–77.

Mansfeld F, Shih H, Postyn A, Devinny J, Islander R, Chin CL (1990). Corrosion monitoring and control in concrete sewer pipes. *CORROSION/1990*, Paper No. 113. Houston, TX: NACE International.

Mansfeld F, Shih H, Postyn A, Devinny J, Islander R, Chen CL (1991). Corrosion monitoring and control in concrete sewer pipes. *Corrosion*, **47**(5): 369–376.

Maruthamuthu S, Saraswathi V, Mani A, Kalyanasundaram RM, Rengaswamy NS (1997). Influence of freshwater heterotrophic bacteria on reinforced concrete. *Biofouling*, **11**(4): 313–323.

Mittleman, NW, Danko JC (1995). Corrosion of a concrete dam structure: evidence of microbially influenced corrosion activity. *International Conference on Microbially Influenced Corrosion*, New Orleans, LA. Welding Society and NACE International, pp. 15-1–15-7.

Molitoris HP (1995). Decomposition of plastics in the sea. *Proceedings of the 9th Int. Congress on Marine Corrosion and Fouling*, Portsmouth, UK.

Moosavi AN, Dawson JL, King RA (1986). The effect of sulphate-reducing bacteria on the corrosion of reinforced concrete. In: Dexter SC (ed) *Biologically Induced Corrosion*. Houston, TX: NACE International, pp. 291–308.

Mori T, Koga M, Hikosaka Y, Nonaka T, Mishina F, Sakai Y, Koizumi J (1991). Microbial corrosion of concrete sewer pipes, H_2S production from sediments and determination of corrosion rate. *Water Sci. Technol.*, **23**: 1275–1282.

Mori T, Nonaka T, Tazaki K, Koga M, Kikosaka Y, Noda S (1992). Interactions of nutrients, moisture, and pH on microbial corrosion of concrete sewer pipes. *Water Res.*, **26**(1): 29–37.

Neu TR, van der Mei HC, Busscher HJ, Dijk F, Verkerke GJ (1993). Biodeterioration of medical-grade silicone rubber used for voice prostheses: a SEM study. *Biomaterials*, **14**(6): 459–464.

Nuclear Regulatory Commision (1984). Examination of failed tendon wires from Fort St. Vrain. Engineering Report No. 50-267, Public Service Company of Denver, Colorado.

Pendrys JP (1989). Biodegration of asphalt cement-20 by aerobic bacteria. *Appl. Environ. Microbiol.*, **55**: 1357.

Perfettini JV, Revertegat E, Langomazino N (1991). Evaluation of cement degradation induced by the metabolic products of two fungal strains. *Experientia*, **47**: 527–533.

Richner P (1996) Untersuchung in zusammenhang mit einem korrosionsschaden. EMPA Report Prufbericht No. 163'669/1, Swiss Federal Laboratories for Material Testing and Research, Dubendorf (German).

Sand W, Bock E (1984). Concrete corrosion in the Hamburg sewer systems. *Environ. Technol. Lett.*, **5**: 517–528.

Sand W, Bock E, White DC (1987). Biotest system for rapid evaluation of concrete resistance to sulfur-oxidizing bacteria. *Mater. Perform.*, **26**(3): 14–17.

Sand W, Dumas T, Marcdargent S (1992). Tests for biogenic sulfuric acid corrosion in a simulation chamber confirm the on-site performance of calcium aluminate-based concretes in sewage applications. In: Basham KD (ed) *Infrastructure: New Materials and Methods of Repair - Proceedings of the Third Materials Engineering Conference.* New York, NY: ASCE, pp. 35–55.

Saricimen H, Maslehuddin M, Allam IM (1987). A case study of deterioration of concrete in a sewage environment in an Arabian Gulf country. *Durability Build. Mater. J.*, **5**: 145–154.

Severyn G (1991). Concrete protection systems for repair of deteriorated infrastructures from MIC. In: Dowling NJ, Mittleman MW, Danko JC (eds) *Microbially Influenced Corrosion and Biodeterioration.* Knoxville, TN: University of Tennessee, pp. 6–23.

Stahl WH, Pessen H (1953). The microbiological degradation of plasticizers: I. growth on esters and alcohols. *Appl. Microbiol.*, **1**(1): 30–35.

Stranger-Johannessen M (1986). The role of microorganisms in the blistering and debonding of corrosion protective organic coatings. *Proceedings of the XVIIIth FATIPEC Congress*, Venezia, Italy, Vol. 3, pp. 1–13.

Stranger-Johannessen M, Norgaard E (1991). Deterioration of anti-corrosive paints by microbial products. *Int. Biodeterior.*, **27**: 157–162.

Sydney R, Esfandi E, Surapaneni S (1996). Control concrete sewer corrosion via the crown spray process. *Water Environ. Res.*, **68**(3): 338–347.

Tator KB (2003). Preventing hydrogen sulfide and microbiologically influenced corrosion in wastewater facilities. *Mater. Perform.*, **42**(7): 32–37.

Upsher FJ (1976). Microbial attack on materials. *Proceeding of the Royal Australian Chemical Institute* 43–44, **6**: 173.

van der Mei HC, van de Belt-Gritter B, Dijk F, Busscher HJ (1996). Initial biodeterioration of silicone rubber by *C. albicans* and *C. tropicalis* strains isolated from voice prostheses. *Cell. Mater.*, **6**(1–3): 157–163.

Vincke E, Monteny J, Beeldens A, De Belie N, Taerwe L, Van Gemert D, Verstraete WH (2000). Recent developments in research on biogenic sulfuric acid attack of concrete. In: Lens P, Hulshoff L (eds) *Environmental Technologies to Treat Sulfur Pollution – Principles and Engineering*. London: International Water Association, pp. 515–541.

Wagner PA, Ray RI, Hart KR, Little BJ (1996). Microbiological degradation of stressed fiber-reinforced polymeric composites. *Mater. Perform.*, **35**(2): 79–82.

Wagner PA, Ray RI, Little BJ, Tucker WC (1994). Microbiologically influenced degradation of fiber reinforced polymeric composites. *Mater. Perform.*, **33**(4): 46–49.

Walch M, Jones-Meehan J (1990). Microbiologically influenced corrosion of epoxy- and nylon-coated steel by mixed microbial communities. *CORROSION/1990*, Paper No. 112. Houston, TX: NACE International.

Chapter 10

Strategies to Prevent or Mitigate Microbiologically Influenced Corrosion

INTRODUCTION

Strategies to mitigate or control the effects of microbiologically influenced corrosion (MIC) include the following: reduce the numbers and types of organisms in the system, use selected bacteria to inhibit corrosion, or alter potential electron acceptors to inhibit specific groups of bacteria.

REDUCE NUMBERS AND TYPES OF ORGANISMS

Numerous methods have been used for minimizing the accumulation of biofilms on engineering surfaces including the following: addition of biocides (oxidizing and nonoxidizing) to the bulk water to kill organisms entering the system or reduce the growth rate of microorganisms within the biofilm, mechanical removal of biofilms from the substratum (sponge balls, brushes), and water treatments to decrease the numbers and types of organisms (aeration and deaeration).

Biocides

Water cleanliness usually refers to turbidity or total suspended solids in the water. Geesey (1993) reported that improving water cleanliness was not necessarily a solution to MIC and that as long as any microorganisms were able to survive, the potential for MIC existed. Geesey (1993) also indicated that while water cleanliness was not effective in controlling MIC, surface cleanliness of piping and equipment was extremely

Microbiologically Influenced Corrosion By Brenda J. Little and Jason S. Lee
Published 2007 by John Wiley & Sons, Inc.

important. Any treatment that regularly physically or chemically cleaned metal surfaces helped prevent or minimize MIC. Mittleman (2002) suggested reducing the amount of assimilable carbon concentration in water to reduce fouling and control MIC.

A biocide is a product formulated to kill microorganisms. Biocides may be applied as batch doses, continuous injections, or a combination of both. Compatibility with equipment, solubility, dose level, dose frequency, chemical compatibility, safety, persistence, toxicity, cost, and solubility influence the selection and application of a biocide. Several excellent reviews have been written about biofilm and MIC control using biocides (Characklis, 1990; Stein, 1993). In all cases, qualification tests are required to ensure that a particular biocide is effective in a particular application. Typical water treatments (Table 10-1) can disinfect down to 10^3 bacterial counts per milliliter (Stein, 1993). It is well established that it is more difficult to kill bacteria in biofilms with biocides than it is to kill the same types of organisms suspended in a liquid medium because biocides cannot penetrate biofilms (Stein, 1993). Costerton et al. (1994) reported that bacteria in biofilms were resistant to antibiotics and biocides at levels 500 to 5000 times higher than those required to kill planktonic cells of the same species. There are additional problems with the use of biocides. Persistent use of a single biocide treatment can allow more resistant microorganisms to develop and remain in the biofilm. Ridgway et al. (1984) demonstrated that bacteria previously exposed to chlorine were more resistant than those never exposed. Resistance to a particular biocide can be overcome by periodically changing the biocide. Numerous investigators have observed a rapid resumption of biofouling after a biocide treatment or mechanical removal of cells from a surface. Regrowth or recovery may be due to the following: (1) Remaining viable cells reproduce a biofilm. (2) Residual biofilm imparts a surface roughness that enhances transport and sorption. (3) Oxidation of extracellular polymeric substances and lysed cells may provide nutrients for regrowth.

Chlorination is a common means of controlling biofilm formation in utilities and in the process industry. Since transport of chlorine to the biofilm is diffusion-controlled, the rate at which chlorine is transported through to the biofilm depends on the concentration of the chlorine in the bulk water and the amount of turbulence. The reaction of chlorine in the bulk water is referred to as the chlorine demand of the water. Chlorine in the water can inactivate microbial cells and oxidize nutrients. Chlorine reacts with organic and reduced inorganic components within the biofilm, disrupting cellular material and inactivating cells. In a mature biofilm, chlorine may react with EPS responsible for the integrity of the biofilm. Biofilms rich in EPS exhibit a more rapid, greater chlorine demand than cells with little associated EPS. Hypochlorite oxidizes EPS within the biofilm, resulting in EPS depolymerization, dissolution, and detachment.

Morris et al. (1995) collected use-level and efficacy information on biocides and corrosion inhibitors from 33 locations within the natural gas industry (Table 10-2). Their descriptions of chemicals and findings are summarized in the following sections.

Glutaraldehyde, an acyclic, 5-carbon compound with terminal carbons constituting aldehyde functional groups, is a nonoxidizing biocide that kills by denaturing cell proteins. Morris et al. (1995) reported that glutaraldehyde was applied in subsurface natural-gas storage and production formations, well tubing and casings,

TABLE 10-1 Typical Water Treatments

Water treatment	Operational limits	Limitations/precautions	Site/system technical data required	Comments
Oxidizing biocides				
Chlorine	pH 6.5 to 7.5, EPA discharge limits	Continuous treatment in addition to slug feed required for bacterial inhibition	Seasonal variations in bacteria content, pH, water/system temperature, suspended solids, organic matter, chlorine demand, ammonia, iron, and manganese levels	Can be used with multiinjection points to maintain TRC
Chlorine dioxide	pH up to 8.7, EPA discharge limits	Gas is extremely volatile and is easily stripped from water systems	Same as chlorine	2.5 times more effective than chlorine for slime control
Hypobromous acid	pH up to 8.7, EPA discharge limits	Continuous treatment in addition to slug feed required for bacterial inhibition	Seasonal variations in bacteria content, pH, water/system temperature, suspended solids, organic matter, chlorine demand, ammonia, iron, and manganese levels	
Hydrogen peroxide	None	Requires 200 ppm dose and long contact time, ~24 h	pH, water/system temperature, suspended solids, organic matter, chlorine demand, ammonia, iron, and manganese levels	Nonpolluting; applicable to stagnant systems
Ozone	Oxides iron and manganese, causing deposition of inorganic material	Plastics/rubber gaskets can become brittle after prolonged exposure to ozone	Seasonal variations in bacteria content, pH, water/system temperature, suspended solids, organic matter, chlorine demand, ammonia, iron, and manganese levels	Applicable to stagnant systems

Continued

239

TABLE 10-1 *Continued*

Water treatment	Operational limits	Limitations/precautions	Site/system technical data required	Comments
Nonoxidizing biocide				
Cationic surfactants	Inactivated by high suspended solids; requires large dosage to be effective	Can produce minor crevice corrosion in stainless steel	Seasonal variations in bacteria content, pH, water/system temperature, suspended solids, organic matter, chlorine demand, ammonia, iron, and manganese levels	Useful in closed and low-flow systems
Nonchemical				
Ultraviolet radiation	Low flow	Requires microscreening (20 μm)	Suspended solids	Can be used in demineralized water and low-flow systems; no by-products

Source: Stein (1993. © NACE International).
Note: Biocides with dispersants may be used in stainless-steel systems. The dispersant can increase the corrosion rate of carbon steel.

TABLE 10-2 Active Ingredients and Number of Cases of Use Reported for Biocides

Active ingredient	Number of cases
Glutaraldehyde	9
Quaternary ammonium compounds (QAC)	3
Acrolein	1
Isothiazolin	1
Diamine acetates (cocodiamine)	1
Carbamates	1
Methylene-bis-thiocyanate	1

Source: Morris et al. (1995. © NACE International).

liquid–gas separators, gathering and transmission lines, and produced water storage and liquid fuel tanks to control bacteria and algae. Glutaraldehyde is soluble in water and insoluble in oils. Glutaraldehyde is incompatible with alkaline substances or strong acids. Morris et al. (1995) investigated the use of glutaraldehyde to treat a sour natural-gas reservoir. In some cases, the treatment was ineffective in reducing sulfide production. The ineffectiveness was attributed to insufficient contact time of the biocide in the treated zone.

Quaternary ammonium compounds (QAC) are ammonium ions (NH_4^+) with alkyl groups substituted for hydrogens. QAC dissolve lipids and cause leakage of vital cell material. Biocides containing QAC may be formulated with bis-tributyl tin oxide (TBT), potassium hydroxide, alcohols, and water. TBT is a broad-spectrum biocide effective for control of bacteria and fungi. Alcohols may provide some biocidal and penetrating capability. As corrosion inhibitors, QAC form protective films on the internal surfaces, reducing exposure to oxidants. Efficacy depends on dose and local conditions. Primary applications are in closed components such as tubing, pipelines, and liquid–gas separators. "Quaternary ammonium compounds are incompatible with strong oxidizing agents such as chlorine, iodine, peroxides, chromates, nitric acid, perchlorates, concentrated oxygen, and permanganates" (Morris, 1995).

Acrolein is an acyclic three-carbon compound with both aldehyde and vinyl functional groups. Biocidal action results from both functional groups. The vinyl group is the most reactive and toxic. Acrolein is an efficient sulfide scavenger reacting with dissolved H_2S and metal sulfides to form complex sulfur compounds. Acrolein is incompatible with amines, bisulfites, sulfites, sulfur oxide, imidazolines, peroxides, caustic alkaline bases, chlorine, and iodine.

Isothiazolin, a cyclic compound containing sulfur, nitrogen, and oxygen, is usually chlorinated, methylated, and soluble in water. It has been applied in a wide range of industrial settings, including water-cooled air-conditioning systems, oil-field injection systems, and drilling muds. Isothiazolin is deactivated by H_2S and is not effective in sour environments.

Diamine acetates (cocodiamine) contain two amine functional groups. In efficacy tests, they were found to be ineffective in controlling MIC bacteria in fluids and on surfaces of undesignated carbon steel at the recommended concentrations (Morris et al., 1995).

Blenkinsopp et al. (1992) demonstrated that the biocidal concentrations of kathon, glutaraldehyde, and QAC were lower when applied within a low-strength electrical field (± 12 V cm^{-1}) with a low current density (± 2.1 mA cm^{-2}). They termed their observation the "bioelectric effect." The bioelectric effect requires a current flow, not just an electric field (Stewart et al., 1999). Subsequent research has confirmed the effect over a range of conditions and biocides. Stewart et al. (1999) suggest that the bioelectric effect is due to increased delivery of oxygen to the biofilm by electrolysis. The mechanism by which oxygen enhances the susceptibility of aerobic microorganisms in biofilms to biocides has not been established. The bioelectric effect has been studied in an effort to prevent and treat device-related bacterial infections and has never been attempted as a method for controlling microbial populations related to MIC.

CORROSION INHIBITION BY BIOFILMS

There are numerous reports of corrosion inhibition by biofilms (Soracco et al., 1984; Pedersen and Hermansson, 1989, 1991; Jayaraman et al., 1997a, b; 1999a, b; Mohanan et al., 1996; Ismail et al., 2002). Corrosion inhibition due to the presence and activities of bacteria within biofilms has been reported for carbon steel (UNS G1018) (Jayaraman et al., 1997b), aluminum 2024 (UNS A92024) (Örnek et al., 2002; Zuo et al., 2005; Jayaraman et al., 1999c), and copper (UNS C10100) (Jayaraman et al., 1999c). The mechanisms most frequently cited for corrosion inhibition by biofilms are the following: (1) The biofilm forms a diffusion barrier to corrosion products, which stifles metal dissolution. (2) Respiring aerobic microorganisms within the biofilm consume oxygen, decreasing the concentration of that reactant at the metal surface. (3) Microorganisms produce metabolic products that act as corrosion inhibitors, for example, siderophores. (4) Microorganisms produce specific antibiotics that prevent the proliferation of corrosion-causing organisms, for example, sulfate-reducing bacteria (SRB). In all the reported cases summarized below, the focus has been on the corrosion-inhibiting properties of specific microorganisms. The reports indicate that biofilms can inhibit both general and localized corrosion.

Several investigators have demonstrated that aerobic bacteria in a biofilm decreased the rate of corrosion of carbon steel. Hernandez et al. (1994) observed increased polarization resistance (R_p) when carbon steel (undesignated, 0.1 percent C, 0.5 percent Mn, 0.04 percent P, 0.04 percent S) was exposed to a synthetic seawater medium [nine-salt solution (NSS)] containing *Pseudomonas* sp. or *Serratia marcescens*. They measured a marked increase in R_p values for the exposures in the presence of either bacterium (Table 10-3). Surface biofilms were stained with acridine orange and imaged with epifluorescence microscopy. Higher R_p values for surfaces colonized by *S. marcescens* compared to those colonized by

Pseudomonas sp. S9 were attributed to higher numbers of *S. marcescens* cells attached to the metal surface. Impedance spectra (Figure 10-1) after immersion in sterile and *Pseudomonas* sp. S9 inoculated NSS demonstrated the same basic trends indicated by the R_p data. The phase angle versus frequency graph showed a maximum at 45° for sterile NSS, indicating Warburg-type impedance and a diffusion-controlled reaction. In contrast, the maximum phase angle in bacterial suspensions was 75° and the modulus versus frequency graph indicated little corrosion during the first 20 days.

TABLE 10-3 R_p of Carbon Steel (kΩ cm^2) After Different Times of Exposure to NSS with and without Bacteria (4×10^8 cells mL^{-1})[a]

Exposure time (days)		
10	20	30
28 ± 3	[b]	[b]
35 ± 5	34 ± 3	15 ± 2
5.0 ± 0.5	4 ± 0.5	3.5 ± 0.5

Source: Hernandez et al. (1994. © NACE International).

[a] Results correspond to a mean value obtained on five specimens $\pm$ the standard deviation.

[b] The system was contaminated after 20 days.

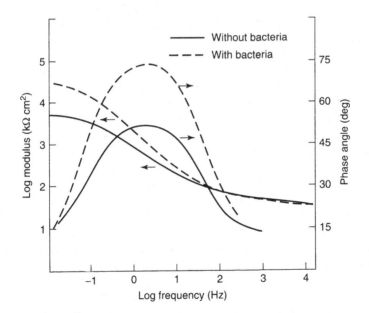

FIGURE 10-1 EIS spectra of an undesignated carbon steel obtained after 20 days of exposure in NSS with and without *Pseudomonas* sp. S9. Curves are given for individual runs. Standard deviation between replicates was ± 10 percent. (Hernandez et al., 1994. © NACE International.)

In addition, the authors determined the following:

(1) Corrosion inhibition required bacterial adhesion.

(2) The inhibition effect disappeared when *in situ* cells were fixed in glutaraldehyde.

(3) When cell-covered carbon steel surfaces were transferred to nutrient-deficient synthetic seawater, the inhibition continued despite the predicted diminished respiration.

(4) After exposure to natural seawater, the inhibitive effect disappeared and *Pseudomonas* could not be located in the biofilm.

Hernandez et al. (1994) speculated that corrosion inhibition might have been due to the formation of a protective layer containing bacteria, exopolymers, and other metabolic materials. In addition, since *Pseudomonas* sp. is an aerobe and *S. marcescens* is a facultative organism that can consume oxygen, the authors concluded that corrosion inhibition could be related to O_2 consumption. Dissolved O_2 concentrations were similar in bulk solutions with and without bacteria. Bacterial concentrations on the metal surface did decrease the number of O_2 molecules that were reduced on the surface. Pedersen and Hermansson (1989, 1991) observed that lowering the oxygen levels of sterile medium by 20 to 25 percent did not reduce corrosion levels of carbon steel (conforming to ASTM A619) to those observed with *Pseudomonas* sp., indicating that respiration alone was not responsible for the corrosion inhibition. Hernandez et al. (1994) hypothesized that at low O_2 partial pressure under biofilms, a passive film of magnetite may have formed that would not have formed without the microorganisms. Mohanan et al. (1996) evaluated corrosion products on carbon steel in naturally occurring pond water. They enumerated heterotrophic and sulfur-oxidizing bacteria and used x-ray diffraction analyses to support the hypothesis that corrosion products formed on undesignated carbon steel in the presence of organic complexes showed better corrosion inhibition than abiotic films.

However, the explanations proposed by Hernandez et al. (1994) are not completely satisfactory since corrosion inhibition was not observed with killed cells but was observed when respiration was reduced in stressed cells. The absence of *Pseudomonas* after a 6-week exposure in natural seawater may indicate competition from other bacteria once the controlled corrosion-inhibiting biofilms are introduced into natural environments.

Jayaraman et al. (1997b) designed experiments to investigate whether or not corrosion inhibition by aerobic biofilms was a general phenomenon. They used G1018 carbon-steel coupons exposed in complex liquid media [Luria Bertani (LB) broth] and Vaatanen nine salt solution (VNSS) and 15 different pure-culture bacterial suspensions representing seven genera. Surface biofilms were stained with a viability assay kit and imaged with a confocal microscope. Compared to sterile controls, the mass loss in the presence of bacteria decreased 2- to 15-fold (Table 10-4). Corrosion inhibition varied among the genera and the extent of corrosion inhibition depended on the nature of the biofilm: An increased proportion of live cells decreased corrosion. Corrosion inhibition was greatest for *Pseudomonas putida*, which produced a more

TABLE 10-4 Bacterial Strains, Antibiotic Resistances, and Corrosion (mg cm^{-3}) of UNS C10180 Carbon Steel Coupon and Standard Deviations Are Shown

Bacterium	Corrosion in LB (mg cm^{-2})	Corrosion in VNSS (mg cm^{-2})	Antibiotic resistance (μg mL^{-1})
Sterile LB medium	1.03 ± 0.14	1.37 ± 0.15	–
Streptomyces lividans TK23.1	0.51 ± 0.08	0.24 ± 0.03	Thiostrepton (50)
Bacillus subtilis	0.39 ± 0.06	Did not grow	–
Bacillus circulans	0.26 ± 0.06	0.37 ± 0.05	–
Rhizobium meliloti 102F34	0.18 ± 0.06	0.17 ± 0.02	–
Pseudomonas fragi K	0.17 ± 0.02	0.52 ± 0.08	Kanamycin (100)
Escherichia coli BK6	0.16 ± 0.04	0.45 ± 0.08	Tetracycline (25)
Bacillus brevis	0.14 ± 0.04	0.19 ± 0.07	–
Burkholderia cepacia G4	0.13 ± 0.04	0.22 ± 0.02	Ampicillin (50)
Agrobacterium tumefaciens A114	0.13 ± 0.03	0.23 ± 0.05	–
Bacillus migulanus	0.11 ± 0.02	0.74 ± 0.07	–
Escherichia coli HB101/pRK2013	0.11 ± 0.02	0.41 ± 0.05	Kanamycin (50)
Pseudomonas mendocina KR1	0.10 ± 0.01	0.14 ± 0.02	Ampicillin (50)
Pseudomonas fluorescens 2-79	0.09 ± 0.01	0.36 ± 0.11	Ampicillin (50)
Pseudomonas putida KT2440	0.09 ± 0.02	0.38 ± 0.05	Ampicillin (50)
Pseudomonas putida F1	0.07 ± 0.01	0.46 ± 0.02	Ampicillin (50)

Source: Jayaraman et al. (1997b). Reprinted with permission of Springer Science and Business Media.

uniform biofilm as determined by confocal scanning laser microscopy than other bacterial species. Corrosion of coupons exposed to *Streptomyces lividans* with only patchy biofilm formation was similar to the sterile controls. Bacterial by-products in the medium did not account for the observations. The authors concluded that a thin layer of actively respiring cells was required to inhibit corrosion.

Iverson (1987) reviewed and summarized the corrosion of a variety of copper alloys in freshwater and seawater. Corrosion was inhibited by the addition of bacteria, but corrosion increased after the bacteria died. Jayaraman et al. (1999c) reported a 20-fold increase in the low-frequency impedance value of C101000 copper as well as a 4- to 7-fold increase in the polarization resistance of aluminum A92024 after 6 days of exposure to axenic aerobic biofilms of either *Pseudomonas fragi* or *Bacillus brevis* in modified Baar's medium. The word "axenic" refers to the growth of organisms of a single species in the absence of cells or living organisms of any other species. The lower corrosion rates were accompanied by increases in corrosion potential (E_{corr}), ennoblement, suggesting the presence of corrosion-inhibiting compounds, according to the authors.

Jayaraman et al. (1999a) used genetically engineered biofilms in a complex, nutrient-rich medium (modified Baar's) to produce antimicrobials specific for SRB. They calculated a 12-fold reduction in the corrosion rate of S30400 stainless-steel when SRB growth was inhibited by *in situ* production of bactenecin.

In addition to electrochemical impedance spectroscopy (EIS) (Figure 10-2a–c), Ismail et al. (2002) measured polarization curves in sterile and inoculated LB medium (Figure 10-3). On the basis of the polarization data, the corrosion rate in sterile medium was 56 and 2 μm year^{-1} in the presence of *Pseudomonas fragi*. The corrosion-controlling effect of the bacteria was accompanied by an ennoblement of E_{corr} by 130 mV. Ismail et al. (2002) also reported a 20-fold reduction in the corrosion of G10180 carbon steel exposed to *P. fragi*. Live cells were necessary for the corrosion reduction, indicating an active role rather than a barrier effect of the biofilm. Flowing nitrogen through the solution was not as effective as *P. fragi* in lowering the corrosion rate. There was an observed ennoblement of the corrosion potential in the presence of *P. fragi* compared with E_{corr} for sterile conditions, suggesting that a metabolic product may be acting as a corrosion inhibitor. Corrosion inhibition increased as medium flow decreased from 12 to 2 mL h^{-1}. The authors interpreted this observation to mean that corrosion inhibition at low flow was due to bacteria in the supernatant and the biofilm, while corrosion inhibition at higher flow rates was due to bacteria only in the biofilm. Additional nutrients did not improve corrosion inhibition.

Eashwar and Maruthamuthu (1995) reviewed the literature on ennoblement (a positive shift in E_{corr}) of metals in marine environments. They concluded that ennoblement is regulated by metabolic activity rather than the physical presence of microorganisms in a biofilm and hypothesized that a strengthening of the passive film on stainless-steel alloys is siderophore-assisted. Siderophores are iron chelators formed by bacteria at near-neutral pH. Others have reported corrosion-inhibition properties of siderophores (McCafferty and McArdle, 1992). Eashwar and Maruthamuthu (1995) rationalized the range of ennoblement that has been reported

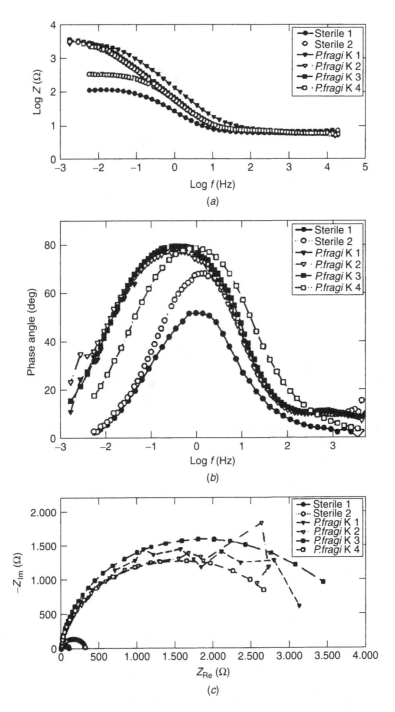

FIGURE 10-2a–c   Electrochemical impedance spectroscopy spectra illustrated as (a) Bode, (b) phase angle versus frequency, and (c) Nyquist plots for G10200 carbon steel after 10 days of exposure in LB medium. (Ismail et al.. 2002. © NACE International.)

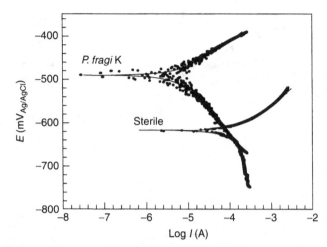

FIGURE 10-3 Polarization curves for G10200 carbon steel in sterile LB medium and in the presence of *P. fragi K* after 15 days, using a scan rate of 0.5 mV s^{-1}. (Ismail et al., 2002. © NACE International.)

in the literature as possible due to alloy–siderophore compatibility, that is, an intrinsic property of the alloy to profit from the presence of the inhibitor.

Örnek et al. (2002) demonstrated that pitting attack of aluminum A92024 in LB medium was reduced by the anionic peptides polyaspartate and polyglutamate, secreted by *Bacillus subtilis* and *Bacillus licheniformis*, respectively. Formation of an aluminum polyaspartate or polyglutamate complex may have reduced the uniform corrosion rate of aluminum.

Despite laboratory studies indicating possibilities for using bacteria to inhibit corrosion of a number of alloys, applications have not been totally successful. Arps et al. (2003) evaluated the concept of corrosion control using regenerative biofilms at Three Mile Island (TMI) Nuclear Power Station, Harrisburg, PA. The field experiment was conducted using a consortium of five bacteria (one polymyxin-producing strain, *Paenibacillus polymyxa* 10401, and four gramicidin S-producing bacillus strains) to inoculate service water in a side stream containing coupons of S30400 stainless steel, G10180 carbon steel, and C26000 cartridge brass under different flow rates. Their results indicate that the inhibition of corrosion, measured by reciprocal polarization resistance ($1/R_p$, instantaneous corrosion rate), by the consortium was small (Figure 10-4a, b). Lower flow rates resulted in higher E_{corr} values. Pitting of C26000 brass specimens was noted in the presence or absence of the consortium (Figure 10-5a, b). Both pit densities and pit areas were lower for samples exposed to the tailored consortium or to the single organisms *P. polymyxa* 10401. Neither $1/R_p$ nor pit area/density provide information as to the extent of localized corrosion.

The following critical issues must be addressed before bacteria can be used to predictably inhibit corrosion: (1) apparent contradictions among researchers, (2) the stochastic nature of biofilms, (3) contamination and natural competition, (4) the influence of nutrients on electrochemical measurements, and (5) the influence of nutrients on the corrosion mechanism.

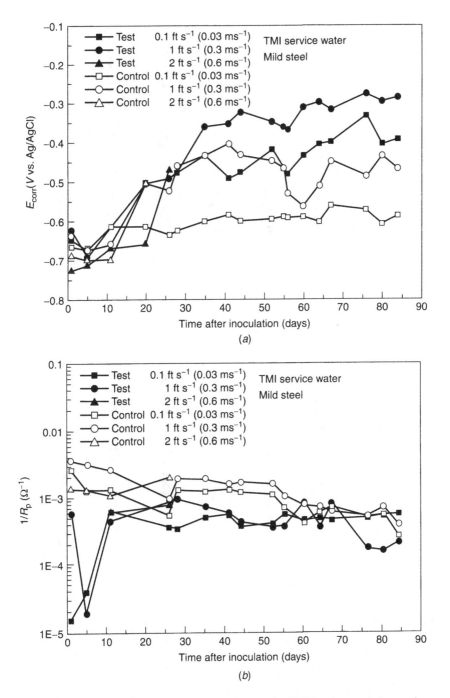

FIGURE 10-4a, b   Time dependence of (*a*) E_{corr} and (*b*) for G1018 carbon steel disc specimens exposed to a bacterial consortium. The test (▲) and control (△) specimens at 2 ft s^{-1} were monitored for about 3 weeks. (Arps et al., 2002. © NACE International.)

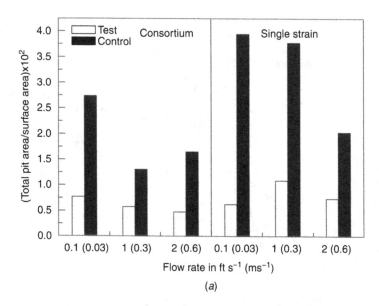

(a)

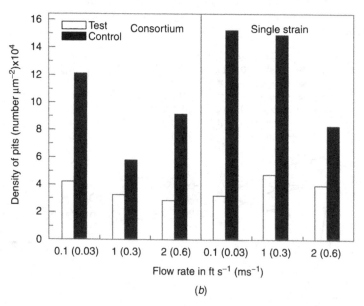

(b)

FIGURE 10-5a, b   Comparative pitting results of C26000 cartridge brass specimens exposed to Three Mile Island service water in the bacterial consortium and single-strain (*Paenibacillus polymyxa* 10401) experiments: (*a*) relative pit area and (*b*) density of pits. Data from nine specimens (a top, side, and bottom specimen for each flow rate) per side stream were used. (Arps et al., 2002. © NACE International.)

Apparent Contradictions among Researchers

Literature on corrosion inhibition by biofilm is confusing because the same organisms and mechanisms, which reportedly cause MIC, can also inhibit corrosion. For example, Pedersen et al. (1988) reported that *Pseudomonas* sp. S9 and *Serratia marcescens* EF 190 caused an increase in the corrosion rate of iron and nickel coated on glass slides compared with sterile conditions. They coated glass slides with a layer of iron and nickel and monitored corrosion as the appearance of transparent patches "at the site of contact between the metal and the bacterial colony." Their experiments were conducted on agar plates containing VNSS and Lewins Marine Medium at 20 °C. The same researchers have shown that *Pseudomonas* sp. S9 and *Serratia marcescens* EF 190 can have a protective effect on carbon steel (ASTM A619) in VNSS at 18-20 °C (Pedersen and Hermansson, 1989,1991). Metal binding by extracellular polymers has been reported as a mechanism for both MIC (Geesey and Jang, 1989) and corrosion inhibition (Ford et al., 1988).

The Stochastic Nature of Biofilms

One of the fundamental assumptions in much corrosion inhibition by biofilm work is that biofilm formation is predictable and controllable. Microorganisms colonize all engineering materials, but there is a stochastic nature to areal coverage and thickness that has never been successfully modeled. Bacteria in pure cultures or in consortia do not form uniform, predictable biofilms. Growth rate depends on substratum, available nutrients, temperature, and electron acceptors. In some of the published papers, the authors have assumed the formation of a uniform, robust biofilm based on prior observations with the test organisms and have not verified the presence, thickness, or areal coverage of the surfaces when interpreting the data. Cells within biofilms die and can cause aggressive corrosion. Biofilm composition is affected by small perturbations in the environment, for example, temperature, nutrient concentration, and flow. The response of microorganisms within the biofilm cannot be predicted with certainty. None of the reports on corrosion inhibition by biofilms has mentioned sloughing, a phenomenon characteristic of all biofilms. Clumps of cells can slough from the surface, transforming a homogeneous biofilm into a patchy one.

Contamination and Natural Competition

Natural competition for extraneous organisms can alter the microbial constituents of a biofilm. Hernandez et al. (1994) observed contamination of controls after brief experiments (20 days) and a change in microbial composition of biofilms after introduction into a natural system. The bigger problem is that most investigators do not evaluate the microbial population at the end of the experiment and have no insight into possible contamination or changes in the engineered biofilm.

The Influence of Nutrients on Electrochemical Measurements

An additional complication is the effect of culture media on electrochemical measurements and on corrosion reactions. Most of the laboratory studies on corrosion inhibition have been conducted in laboratories with nutrient-rich media. The impact of nutrients on electrochemical measurements and MIC has been investigated by Webster and Newman (1994). They observed interferences in electrochemical measurements when yeast extract was included in the culture medium and electrolyte. The interferences were removed when the yeast extract was removed. All the nutrient-rich media used in laboratory studies contained yeast extract. Modified Baar's medium contains (in g L^{-1}) 10 tryptone, 5 yeast extract, and 10 NaCl. The Luria–Bertani (LB) medium contains (in g L^{-1}) 10 tryptone, 5 yeast extract, and 10 NaCl. Vaatanen nine-salt solution consists of 1.0 g peptone, 0.5 g glucose, 0.5 g starch, 0.5 g yeast extract, 0.01 g hydrated ferrous sulfate, 0.01 g NaHPO$_4$, and 1000 mL of NNS containing 17.6 g of NaCl. Lewins Marine medium consists of 5 g yeast extract, 5 g tryptone, 1 g tris (hydroxymethyl) aminomethane, 0.1 g sodium-glycero-phosphate, and 1000 mL of NNS. The cited studies on corrosion inhibition in complex media have compared corrosion in sterile and inoculated media, but none has carefully studied the impact of media constituents on the electrochemical parameters used to quantify corrosion.

The Influence of Nutrients on the Corrosion Mechanism

Webster and Newman (1994) examined the impact of media constituents on localized corrosion and made the following observations: Localized corrosion would not readily occur unless chloride ion was the predominant anion in the medium. They concluded that chloride must be present in a concentration at least comparable to that of all other anions combined, otherwise corrosion was inhibited even at high H$_2$S concentrations of up to 500 ppm. Reduction of the ratio of Cl$^-$ ions to other inhibiting anions increased the time to initiation and decreased the rate of propagation of the corrosion. Other corrosion investigators have concluded that extra nutrients cannot be added to stimulate bacterial growth if those nutrients inhibit corrosion by adding too many nonchloride ions (Ringas and Robinson, 1988). Anions, including sulfate, hydroxide, phosphate, acetate, carbonate, and nitrate can inhibit pitting corrosion. It is possible that bacterial consumption and fixation of nutrients, including sulfate, could render an initially inhibiting solution aggressive by removing nonchloride ions (see Chapter 2 for further discussion).

Whereas most groups studying corrosion inhibition by biofilms focused on the organisms within biofilms, groups at the State Research Center for Microbiology, Moscow, Russian Federation (Jigletsova et al., 2004; Rodin et al., 2005) carefully examined the influence of environmental conditions on corrosion inhibition by biofilms. They demonstrated that the division of bacteria into ones that caused corrosion and ones that inhibited corrosion was entirely arbitrary. They demonstrated

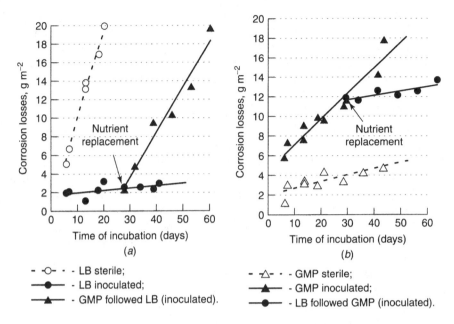

- ⋅-○-⋅⋅ - LB sterile;
- ━●━ - LB inoculated;
- ━▲━ - GMP followed LB (inoculated).

- ⋅-△-⋅⋅ - GMP sterile;
- ━▲━ - GMP inoculated;
- ━●━ - LB followed GMP (inoculated).

FIGURE 10-6a, b   The dynamics of corrosion losses with nutrient replacement: (a) the biofilm grown in LB was transferred into GMP and (b) the biofilm grown in GMP was transferred into LB medium. (Rodin et al., 2005. © NACE International.)

that the corrosive properties of biofilms varied with culture conditions. They used undesignated carbon steel coupons exposed to a natural consortium of bacteria isolated from oil-processing waters. The organisms included oil-oxidizing aerobes and SRB. An increase in corrosion was observed when the coupon was transferred from LB to glucose-mineral medium supplemented with peptone (GMP) (Figure 10-6a). During biofilm formation in a GMP, corrosion losses increased versus sterile control. However, corrosion decreased when coupons with biofilms were transferred into enriched LB medium (Figure 10-6b). Their data indicate that environmental conditions determine the specific microbiological effect on corrosion processes, not the individual organisms. Dubiel et al. (2002) used electrochemical impedance spectroscopy to evaluate corrosion inhibition in the presence of iron-reducing bacteria. They concluded that the physiology of the bacteria in the biofilm, the flow rate, and the chemistry of the electrolyte determine the ultimate impact of microorganisms on corrosion.

ALTER POTENTIAL ELECTRON ACCEPTORS TO INHIBIT SPECIFIC GROUPS OF BACTERIA

One practical application for controlling MIC by controlling the electrolyte composition has been used in seawater injection systems. In these systems, seawater is injected into oil reservoirs to maintain pressure. Oxygen is removed to minimize corrosion. However, in the anaerobic environment, growth of SRB is encouraged and corrosion of iron and steel alloys is the result. Laboratory and field experiments have demonstrated that nitrate treatment can be an effective alternative to biocide treatment to reduce the numbers of SRB and their activity. Nitrate treatment was implemented on an oil platform in the North Sea (Veslefrikk) (Thorstenson et al., 2002). The change from glutaraldehyde treatment to nitrate resulted in a dramatic change in the bacterial community. The SRB population decreased and the numbers of nitrate-reducing bacteria (NRB) increased (Figure 10-7). After 4 months of nitrate addition, the activity of SRB in the biofilm was markedly reduced as measured with respiratory methods and an enrichment of NRB was measured. After 32 months of nitrate treatment, SRB numbers were reduced 20,000-fold and SRB activity was reduced 50-fold (Figure 10-7). Corrosion measurements of an undesignated carbon steel decreased from 0.7 to 0.2 mm year^{-1}. Similar applications have been made to reduce souring (Larsen, 2002; Larsen et al., 2004). Gullfaks platforms (Sunde et al., 2004) have been treated with nitrate to reduce H_2S production. A 1000-fold reduction in SRB numbers, a 10- to 20-fold reduction in sulfate respiration activity, and a 50-percent reduction in corrosion of an undesignated carbon steel were observed. The authors suggest that reservoir characteristics and nutrient availability have a significant impact on the effectiveness of nitrate injection.

Nitrate-reducing bacteria reduce nitrate to N_2 with several possible intermediates, including nitrite. There are several potential mechanisms for the observed inhibition of SRB due to addition of nitrate. One of them is competition for carbon sources. When competing for the same carbon source, NRB out-compete SRB because nitrate is a stronger oxidizer than sulfate. This argument is valid only in carbon-limited waters. Toxic reaction products from the reduction of nitrate to N_2 may inhibit SRB. A shift in the redox potential in the system may also inhibit SRB. As a consequence of nitrate reduction, the redox potential will likely increase, producing unfavorable conditions for sulfate reduction.

Hubert et al. (2004) demonstrated that both nitrate and nitrite are effective treatments for decreasing sulfide concentrations (Table 10-5). The required dose depends on the concentration of oil organics used as the energy source by the microbial community. Because of its higher oxidative power, nitrate can remove more oxidizable oil organics than nitrite. However, nitrite is a strong inhibitor of SRB. Jhobalia et al. (2005) demonstrated that high sulfate concentration in the medium (increases from 1.93 to 6.5 g L^{-1}) could inhibit the sulfate-reduction rate of *Desulfovibrio desulfuricans*.

Voordouw et al. (2002) demonstrated a nitrate-reducing, sulfide-oxidizing bacterium capable of reducing nitrate to nitrite, nitrous oxide, or nitrogen and oxidizing sulfide to sulfate or sulfur. The stoichiometry of the reactions catalyzed by the organism depends on the ratio of sulfide to nitrate. Dunsmore et al. (2004) isolated an organism from a Danish North Sea oil-field water injection system that had been continuously treated with nitrate since the start of the injection. This species, an SRB, could reduce nitrate and produce ammonium in the presence of sulfate, increasing the likelihood of corrosion. Despite the potential, no additional corrosion was measured in the presence of this unique type of respiration.

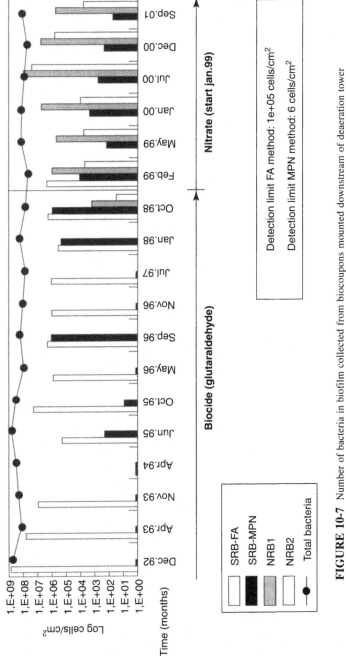

FIGURE 10-7 Number of bacteria in biofilm collected from biocoupons mounted downstream of deaeration tower on a floating oil platform in the North Sea (Veslefrikk). Column marked NRB1 shows number of cells targeting facultatively anaerobic NRB and columns marked NRB and NRB2 show number of cells targeting obligatory anaerobic NRB. FA, fluorescent antibody; MPN, most probable number. Columns in the figure are ordered left to right corresponding to the legend top to bottom. (Thorstenson et al., 2002. © NACE International.)

TABLE 10-5 Weight Loss of ASTM A366 Carbon Steel Coupons Buried in the Sand Matrix of Bioreactors in Response to Nitrate or Nitrate Treatment

Treatment	Weight loss (%)[a]
None	8.3 ± 5.3
Nitrate (17.5 mM)	4.7 ± 1.3
Nitrate (20 mM)	0.7 ± 0.2
Control[b]	0.8 ± 0.1

Source: Hubert et al. (2004. © NACE International).
[a]Average for 15 corrosion coupons (± SD). Total incubation time ~110 days.
[b]Incubated in the absence of nitrate or nitrite for 45 days (batchwise and increasing flow rate periods only).

Hubert et al. (2006) suggested that bioaugmentation, in which *ex situ* grown microorganisms could be injected with the nitrate if indigenous NRB were lacking. Despite the possibility of bioaugmentation, there are several reports of failures. Bouchez et al. (2002) attempted to inoculate a nitrifying sequencing batch reactor with an aerobic denitrifying bacterium. The added bacterium disappeared after two days.

It was once thought that sparging a system with air could alleviate corrosion problems due to anaerobes such as SRB. This has since been disproved. Because SRB depend on other organisms to remove oxygen and produce nutrients, they can survive in aerated systems. Aeration will not reduce the impact of SRB and in some cases exacerbates the problem. Altering electron acceptors can also increase the likelihood of MIC.

Removing oxygen from seawater has been proposed as a corrosion-control measure. Matsuda et al. (1999) conducted shipboard trials by sealing a ballast tank at the deck and installing vertical pipes into the headspace. They reported that pumping pure nitrogen gas into the headspace for 1.5 h reduced oxygen levels in the seawater to approximately 0.2 mg L^{-1} and decreased the rate of uniform corrosion of carbon steels (carbon content between 0.1 and 0.2 percent) by 90 percent as determined by weight loss. However, in laboratory experiments, Lee et al. (2004) compared corrosion of G1020 resulting from stagnant aerobic natural seawater with corrosion resulting from stagnant anaerobic natural seawater over a 1-year period. They demonstrated the following: (1) Corrosion was more aggressive under totally anaerobic conditions as measured by instantaneous corrosion rates ($1/R_p$) and weight loss. (2) Under aerobic conditions, corrosion was uniform and the surface was covered with iron oxides (lepidocrocite and goethite). (3) Under anaerobic conditions, the corrosion was localized pitting and the corrosion products were mackinawite and pyrrhotite. Lee et al. (2005) designed field experiments to evaluate deoxygenation of natural seawater as a corrosion-control measure for unprotected G1020 carbon steel seawater ballast tanks. They demonstrated the difficulty of maintaining hypoxic seawater. With use of a gas mixture, it was possible to displace dissolved oxygen. However, aerobic respiration and corrosion reactions consumed oxygen and produced totally anaerobic conditions within the first days of hypoxia. When gaskets and seals failed, oxygen was inadvertently introduced. The impact of oxygen ingress on corrosion depends on the amount of oxygen in the system at the time oxygen is introduced. G1020 carbon steel exposed to cycles of hypoxic seawater and oxygenated atmosphere had higher corrosion rates than coupons exposed to cycles of either consistently aerobic or deoxygenated conditions. Commercial technologies are available for the deoxygenation of seawater for ballast tanks. There are no data on the long-term benefits of these treatments on corrosion inhibition.

SUMMARY

Biocides control MIC by decreasing the microbial population, whereas control by manipulation of electron acceptor and by corrosion-inhibiting biofilms relies on stimulation or retardation of specific microbial populations. The problems associated with biocide use are well documented and include the following: (1) inability of biocides to penetrate biofilms and (2) rapid regrowth after a biocide treatment.

Corrosion inhibition due to biofilms has been demonstrated in the laboratory for several microorganisms on several metals and alloys, but has never been demonstrated in a field application. The laboratory studies are difficult to compare because of the differing experimental conditions, organisms and culture conditions. The following critical issues must be addressed before bacteria can be used to predictably inhibit corrosion: (1) the stochastic nature of biofilms and (2) contamination and/or natural competition. Controlling MIC by controlling the sulfate and nitrate concentrations has been used successfully in limited applications in the seawater injection systems. The long-term consequences are not known.

REFERENCES

Arps PJ, Xu LC, Green RM, Wood TK, Mansfeld FB, Syrett BC, Earthman JC (2003). Field evaluation of corrosion control using regenerative biofilms (CCURB). *CORROSION/2003*, Paper No. 03714. Houston, TX: NACE International.

Blenkinsopp SA, Khoury AE, Costerton JW (1992). Electrical enhancement of biocide efficacy against *Pseudomonas aeruginosa* biofilms. *Appl. Environ. Microbiol.*, **50**(11): 3770–3773.

Bouchez T, Patureau D, Dabert p, Juretschko S, Dore J, Delgenes P, Molette R (2000). Ecological study of a bioaugmentation failure. *Environ. Microbiol.*, **2**(2): 179–190.

Characklis WG (1990). Microbial biofouling control. In: Characklis WG, Marshall KC (eds) *Biofilms*. New York, NY: Wiley, pp. 585–633.

Costerton JW, Ellis B, Lam K, Johnson F, Khoury AE (1994). Mechanisms of electrical enhancement of efficacy of antibiotics in killing biofilm bacteria. *Antimicrob. Agents Chemother.*, **38**(12): 2803–2809.

Dubiel M, Hsu CH, Chien CC, Mansfeld F, Newman DK. (2002) Microbial iron respiration can protect steel from corrosion. *Appl. Environ. Microbiol.*, **68**(3): 1440–1445.

Dunsmore BC, Whitfield TB, Lawson PA, Collins MD (2004). Corrosion by sulfate-reducing bacteria that utilize nitrate. *CORROSION/2004*, Paper No. 04763. Houston, TX: NACE International.

Eashwar M, Maruthamuthu S (1995). Mechanisms of biologically produced ennoblement: ecological perspectives and a hypothetical model. *Biofouling*, **8**: 203–213.

Ford T, Maki JS, Mitchell R (1988). Involvement of bacterial exopolymers in biodeterioration of metals. In: Houghton D, Smith RN, Eggins HO (eds) *Biodeterioration 7*. Cambridge, UK: Elsevier Applied Science, The Biodeterioration Society, pp. 378–384.

Geesey G (1993). Biofilm formation. In: Kobrin G (ed) *A Practical Manual on Microbiologically Influenced Corrosion*, Vol. 1. Houston, TX: NACE International, pp. 11–13.

Geesey G, Jang L (1989). Binding of metal ions by extracellular polymer. In: Beveridge T, Doyle R (eds) *Metals, Ions and Bacteria*. New York, NY: Wiley, p. 325.

Hernandez G, Kucera V, Thierry D, Pedersen A, Hermansson M (1994). Corrosion inhibition of steel by bacteria. *Corrosion*, **50**(8): 603.

Hubert C, Voordouw G, Nemati M, Jenneman G (2004). Is souring and corrosion by sulfate-reducing bacteria in oil fields reduced more efficiently by nitrate or by nitrite? *CORROSION/2004*, Paper No. 04762. Houston, TX: NACE International.

Hubert C, Voordouw G, Arensdorf J and Jenneman G. (2006). Control of souring through a novel class of bacteria that oxidize sulfide as well as oil organics with nitrate. *CORROSION/2006*, Paper No. 06669. Houston, TX: NACE International.

Ismail Kh M, Gehrig T, Jayaraman A, Wood TK, Trandem K, Arps PJ, Earthman JC (2002). Corrosion control of mild steel by aerobic bacteria under continuous flow conditions. *Corrosion*, **58**(5): 417–423.

Iverson WP. (1987) Microbial corrosion of metals. *Adv. Appl. Microbiol.*, **32**: 1–36.

Jayaraman A, Cheng ET, Earthman JC, Wood TK (1997a). Importance of biofilms formation for corrosion inhibition of SAE 1018 steel by axenic aerobic biofilms. *J. Ind. Microbiol. Biotechnol.*, **18**: 396.

Jayaraman A, Earthman JC, Wood YK (1997b). Corrosion inhibition by aerobic biofilms on SAE 1018 steel. *Appl. Microbiol. Biotechnol.*, **47**: 62.

Jayaraman A, Hallock PJ, Carson RM, Lee CC, Mansfeld FB, Wood TK (1999a). Inhibiting sulfate-reducing bacteria in biofilms on steel with antimicrobial peptides generated in situ. *Appl. Microbiol. Biotechnol.*, **52**: 267–275.

Jayaraman A, Lee CC, Chen MW, Mansfeld F, Wood TK (1999b). Inhibiting sulfate-reducing bacteria in biofilms by expressing the antimicrobial peptides indolicidin and bactenecin. *J. Ind. Microbiol. Biotechnol.*, **22**: 168.

Jayaraman A, Örnek D, Duarte DA, Lee CC, Mansfeld FB, Wood TK (1999c). Axenic aerobic biofilms inhibit corrosion of copper and aluminum. *Appl. Microbiol. Biotechnol.*, **52**: 787–790.

Jhobalia CM, Hu A, Gu T, Nesic S (2005). Biochemical engineering approaches to MIC. *CORROSION/2005*, Paper No. 05500. Houston, TX: NACE International.

Jigletsova SK, Rodin VB, Zhirkova NA, Alexandrova NV, Kholodenko VP (2004). Influence of nutrient medium composition on the direction of microbiologically influenced corrosion. *CORROSION/2004*, Paper No. 04575. Houston, TX: NACE International.

Larsen J (2002). Downhole nitrate applications to control sulfate reducing bacteria activity and reservoir souring. *CORROSION/2002*, Paper No. 02025. Houston, TX: NACE International.

Larsen J, Malene RH, Zwolle S (2004). Prevention of reservoir souring in the Halfdan field by nitrate injection. *CORROSION/2004*, Paper No. 04761. Houston, TX: NACE International.

Lee JS, Ray RI, Lemieux E, Falster A, Little BJ (2004). An evaluation of carbon steel corrosion under stagnant sweater conditions. *Biofouling*, **20**(4/5): 237–247.

Lee JS, Ray RI, Lemieux E, Little, BJ (2005). An evaluation of deoxygenation as a corrosion control measure for ballast tanks. *J. Corrosion*, **65**(12): 1173–1188.

Matsuda M, Kobayashi S, Miyuki H, Yosida S (1999). *An Anticorrosion Method for Ballast Tanks using Nitrogen Gas*. Report of Research and Development to the Ship and Ocean Foundation (Japan), October.

McCafferty E, McArdle JV (1992). Corrosion inhibition by biological siderphore. *Proceedings of the 182nd Society Meeting*, The Electrochemical Society, Toronto, pp. 185–186.

Mittleman M (2002). *MIC of Sprinkler Piping*. ISH North America, International Trade Fair for Kitchen and Bath Plumbing, Heating and Air Conditioning, p. 5.

Mohanan S, Maruthamuthu S, Mani A, Venkatachari G (1996). Corrosion control by biofilm formation. *Anti-Corrosion Meth. Mater.*, **43**(5): 23–27.

Morris EA, Pope DH, Fillow JP, Brandon DM, Fetsko ME, Fulton JW (1995). Current and future trends in biocide and corrosion inhibitor usage in the natural gas industry: efficacy and potential environmental impact. *International Conference on Microbially Influenced Corrosion*. Houston, TX: NACE International, pp. 51/1–51/13.

Örnek D, Jayaraman A, Syrett B, Hsu C-H, Wood TK (2002). Pitting corrosion inhibition of aluminum 2024 by *bacillus* biofilms secreting polyaspartate or γ-polyglutamate. *Appl. Microbiol. Biotechnol.*, **58**: 651–657.

Pedersen A, Hermansson M (1989). The effects on metal corrosion by *Serratia marcescens* and a *Pseudomonas* sp. *Biofouling*, **1**(4): 313.

Pedersen A, Hermansson M (1991). Inhibition of metal corrosion by bacteria. *Biofouling*, **3**: 1.

Pedersen A, Kjelleberg S, Hermansson M (1988). A screening method for bacterial corrosion of metals. *J. Microbiol. Meth.*, **8**: 191.

Ringas C, Robinson F (1988). Corrosion of stainless steel by sulfate-reducing bacteria-electrochemical techniques. *Corrosion*, **44**: 386.

Ridgway HF, Justice CA, Whittaker C, Argo DG, Olson BH (1984). Biofilm fouling of RO membranes – its nature and effect on treatment of water for reuse. *J. AWWA*, **76**(94): 274–287.

Rodin VB, Zhigletsova SK, Zhirkova NA, Alexandrova NV, Shtuchnaja GV, Chugunov VA, Kholodenko VP (2005). Altering enviromental composition as a potential method for reversing microbiologically influenced corrosion. *CORROSION/2005*, Paper No. 05498. Houston, TX: NACE

Soracco RJ, Berger LR, Berger JA Mayack LA, Pope DH, Wilde EW (1984). Microbiologically mediated reduction in the pitting of mild steel overlaid with plywood. *CORROSION/1984*, Paper No. 98. Houston, TX: NACE International.

Stein A (1993). MIC treatment and prevention. In: Kobrin G (ed) *A Practical Manual on Microbiologically Influenced Corrosion*, Vol. 1. Houston, TX: NACE International, pp. 101–112.

Stewart PS, Wattanakaroon W, Goodrum L, Fortun SM, McLeod BR (1999). Electrolytic generation of oxygen partially explains electrical enhancement of tobramycin efficacy against *Pseudomonas aeruginosa* biofilm. *Antimicrob. Agents Chemother.*, **43**(2): 292–296.

Sunde E, Lillebo B-LP, Bodtker G, Torsvik T, Thorstenson T (2004). H_2S inhibition by nitrate injection on the Gullfaks field. *CORROSION/2004*, Paper No. 04760. Houston, TX: NACE International.

Thorstenson T, Bodtker G, Lillebo B-LP, Torsvik T, Sunde E, Beeder J (2002). Biocide replacement by nitrate in sea water injection systems. *CORROSION/2002*, Paper No. 02033. Houston, TX: NACE International.

Voordouw G, Nemati M, Jenneman GE (2002). Use of nitrate-reducing, sulfide-oxidizing bacteria to reduce souring in oil fields: interactions with SRB and effects on corrosion. *CORROSION/2002*, Paper No. 02034. Houston, TX: NACE International.

Webster BJ, Newman RC (1994). Producing rapid sulfate-reducing bacteria-influenced corrosion in the laboratory. In: Kearns J, Little BJ (eds) *Microbiologically Influenced Corrosion Testing*. . Philadelphia, PA: ASTM Publication STP 1232, pp. 28–41.

Zuo R, Kus E, Mansfeld F, Wood TK (2005). The importance of live biofilms in corrosion protection. *Corrosion Sci.*, **47**: 279–287.

Index

Abiotic
  corrosion, 37
  environments, 98
  tests, 114, 116
Acanthite, 31, 39–40
Accelerated low water corrosion (ALWC), 164–169
Accelerating corrosion, 175
Accidents, liquid pipelines, 195–197
Accumulation, 8–10, 45, 103–104, 193
Acetate, 57, 252
Acetic acid, 18, 41, 233
Acetogenic bacteria, biofilm formation, 17–18
Acetone, 71
Acid(s), 6, 25, 41, 217, 219, 227–228
Acidification, 25
Acidithiobacillus spp., 23
Acidity, significance of, 175
Acid-producing bacteria (APB), 57, 66
Ac impedance spectroscopy, 113
Acinetobacter, 232
Acoustic emission (AE), 223, 226
Acroline, 241
Acrylics, 219
Adenosine-5′-phosphosulfate (APS) reductase, 58–59
Adenosine triphosphate (ATP), 58, 71
Adhesion, 7, 9, 12, 138–139, 141, 244
Adhesives, 219
Admiralty brass, 174
Adsorption, 9, 12, 14, 25, 49
Aeration, 82, 129–130, 140, 148
Aerobic
  bacteria, *see* Bacteria, aerobic
  conditions, 23–25, 256–257
  respiration, 5, 28

Afipia, 61
Aggressive corrosion, 49, 175, 256
Agrobacterium tumefaciens, 245
Air
  bumping, 149
  in concrete, 229
  injection, 229
Airborne contaminants, 188
Aircraft, *see* Jet fuels
  fungal growth, 155–157
  microbial contamination, 195, 200–202
Alcaligenes, 232
Alcohols, 5–6, 192
Aldehydes, 192
Algae, 2, 17, 28, 160
Aliphatic
  acids, 192
  amines, 49
Alkaline phosphatase, 63
Alkalines, 227
Alkly-aryl phosphates, 203
Alkyl phosphates, 203
All-epoxy-polymide coatings, 121
Alloying elements, impact of
  aluminum/aluminum alloys, 139–140
  antimicrobial metals, 141–143
  cathodically protected, 169–173
  copper alloys, 128–133
  corrosion-resistant, generally, 173
  low alloy steel, 128–129
  nickel alloys, 128–133
  overview, 127–128, 143
  passive, 173
  stainless steels, 128, 133–139
  titanium/titanium alloys, 140
Alteromonas atlantica, 60

Microbiologically Influenced Corrosion By Brenda J. Little and Jason S. Lee
Published 2007 by John Wiley & Sons, Inc.